Livestock Guardians

How to Use Llamas, Donkeys, Dogs, and More to Safeguard Your Herd and Property

Table of Contents

Introduction

Predators can quickly become a farmer's nightmare. Losing an animal here and there adds up, never mind the devastating blows when an unwelcomed intruder ravages through the majority of your flock. Farmers are fighting an eternal battle against predators, and one of the most humane solutions is enlisting the help of a guardian animal.

Pictures of the galloping guard dog across idyllic hills bring nostalgic visions of simpler times. Dogs or cats are always the first to come to mind, and with good reason, because they are so effective. However, numerous animal species make great guardians, including donkeys and llamas. Furthermore, there are a multitude of dog breeds to choose from, so deciding on the best one could leave your head spinning. Moreover, dogs can quickly devolve into intruders, causing the most significant risks for your livestock if they are not trained and cared for thoughtfully.

This book provides you with all the information you need about guardian animals. This includes the ethical and legal considerations – and which breeds, species, and conditions are best suited for the homestead, the farm, or the smallholding you are building. Choosing a guardian animal blind can be disorienting. Therefore, the guiding light of theoretical information and practical tips is presented to lead you to a successful farm.

Some methods of predator control, like poisoning or trapping, can be inhumane and cause undue suffering. Poison can make the environment around you toxic, and it will kill animals that it was not meant to kill, while traps can keep a predator in prolonged pain. Although livestock guardians attack and kill predators, they are a more humane option because they keep predators away, which will maintain their population in the wild.

Man, meeting nature does not have to be disastrous. By creating a cohesive system that includes your guard animals, you can build an ethical and productive farm that meets and surpasses regulatory standards and is eco-friendly. The bonds you create between your flock and guard animals are the mutually beneficial relationships that highlight how humans and nature can coexist productively. The fulfilling journey of setting up your farm in a way that keeps your animals safe and maximizes their well-being will reward you both mentally and financially.

Following proper guidance lets you rest easy knowing your animals are protected. Acquiring the skills and knowledge this book provides you will make you a much better farmer. From medical considerations to nutrition and habitat, all the bases are covered to make you the best guardian animal owner. Do not hesitate. Dive into this book and unlock your full potential as a livestock keeper. You are your animals' first line of protection, but that does not mean you don't need some assistance along the way. The perfect helping paws or hooves are right at your fingertips if you are curious enough to to go deeper into the world of livestock guardians!

Chapter 1: The Guardian's Role

This chapter discusses the essential role of guardian animals in protecting livestock and property and how they contribute to overall security. You'll find a brief overview of the historical use of guard animals, which essentially started with the domestication of wolves and the initiation of dogs into police and military systems. You'll explore the evolution of guard animals' role over time, discovering the advantages of this predator management method. Finally, you'll learn about ethical and legal considerations regarding the use of guard animals.

The Historical Use and Evolution of Guardian Animals

Around 40,000 years ago, humans noticed incredible potential in wolves. This animal is strong, intelligent, and able to hunt. They realized that wolves would become more trainable, friendly, and receptive to humans by domesticating them. This would allow the domesticated animal to use its territorial and hunting instincts to guard humans and keep their belongings safe. Like modern dogs, wolves have always used barking to warn their pack about imminent danger.

Around 40,000 years ago, man domesticated wolves.
Internet Archive Book Images, CC0, via Wikimedia Commons.
https://commons.wikimedia.org/wiki/File:Image_from_page_82_of_%22Natural_history%22_(18 97)_(20734677882).jpg

Early research suggests that humans separated wolf pups from their packs and trained them to do several tasks. However, this theory is unlikely because the animal is virtually untrainable. Unless they managed to socialize the pups when they were as little as 19 days old, the animals would've never responded to them. Training a young animal is also taxing; it's unlikely that hunter-gatherer early humans – whose only goal was survival – would've taken the time needed for this process.

Another theory was that food waste and other paraphernalia left around humans' early settlements attracted wolves to the area. Wolves are naturally inclined to scavenge, hunt, and look around for food. While these settlements would've undoubtedly made a desirable food source, wolves are also naturally scared of humans and potential sources of danger.

Whenever a wild animal senses the presence of a human or another potentially dangerous entity, their "flight distance" instinct kicks in. This instinct refers to each animal's instinct to allow a source of perceived danger to approach them before they flee. It is reasonable to believe that wolves with weak flight distances were eventually domesticated since having a weak flight distance means they are more sociable than their counterparts. They were more likely to engage with humans and spend more time physically near them. While this theory is unconfirmed, it has been supported by scientific experiments regarding the domestication of foxes.

Mesopotamian Dogs

Regardless of the technicalities, dogs, as everyone knows them today, exist because humans in different corners of the world recognized the potential of wolves. Proof of humans using dogs as guardians goes back to ancient Mesopotamia. Native legend and folklore tell about large animal breeds guarding local livestock. Not only does Mesopotamian art feature dogs abundantly, but they're also generally known to have protected homes and livestock from predators, especially smaller wolves. Dogs were also carved into protective amulets.

The Revered Dogs of Ancient Egypt

Dogs were also sacred to the ancient Egyptians. They held religious significance because they were associated with Anubis, the god of the dead. Anubis was depicted as having a human's body and a dog's head. Dogs were also commonly kept as pets because of their protective nature and hunting abilities, and they were also a great source of companionship. Ancient Egyptians formally mourned the passing of their dogs, which further signifies their reverence.

The First Official Use of Guardian Dogs

Ancient Greek mythology is perhaps the most solid proof of the protective role of dogs at the time. According to lore, Hades, the god of the underworld, employed Cerberus, a three-headed dog, to guard the gates of his realm. The world-renowned Athenian philosopher Plato also frequently employed dogs as philosophical examples. He highlighted their natural roles as hunters and loyal protectors and explained in his work, *The Republic*, that intelligent animals can easily distinguish between friends and foes.

The Molossians, the ancient Greek tribe that lived in Epirus, are known for their role in breeding Molossus dogs. The Molossus breed was used to guard pastures and homes, and while it is now extinct, it's the ancestor of the Mastiff breed. The dog breed also served as a personal guardian and was even sometimes deployed on the battlefield alongside warriors.

Roman Molossers

Ancient Romans also commonly used guard dogs to protect their homes and belongings. Archeologists found evidence in the ancient city of Pompeii hinting at the role of dogs in the Roman Empire. One artifact was a mosaic artwork featuring a massive black dog and text translated into "Beware of the Dog." Similar warning pieces, along with the remnants of a dog that was chained to a temple, were also found.

As the Roman Empire expanded and cultural practices spread to different parts of the world, the idea of using guardian dogs also became more popular. Romans introduced Molossers to the areas they conquered, which led to breeding more Mastiff dogs around the region. At this point in time, however, people didn't know the difference between different dog breeds.

The German Rottweiler

Having a dangerous reputation and being notorious for their incredibly powerful bites, German Rottweilers are among the most popular guardian dogs today. They were among the earliest guard dog breeds to ever exist. They were historically utilized to herd livestock and protect their owners' prized belongings. While it is considered a dangerous breed, the Rottweiler makes a good guardian animal because it can form bonds and companionships with humans and other animals, can be controlled, and is highly intelligent.

German Rottweilers are among the most popular guardian dogs today.
mar_qs, CC0, via Wikimedia Commons: https://commons.wikimedia.org/wiki/File:Rottweiler_-52773841920.jpg

Similar dog breeds were also commonly used to protect homes and property. The Italians also trained the Cane Corso breed to protect and herd livestock and hunt predators. Molosser-type dogs were generally used for entertainment, such as dogfighting and bullbaiting. While these activities are highly unethical, they highlight traits such as controlled aggression and tenacity, which are necessary in guarding animals.

The Emergence of More Dog Breeds

During the 19th century, more dog breeds, including the German Shepherd, emerged. Scientists and breeders were breeding dogs to develop breeds that ultimately carried all the traits they desired within a dog. Some strove to create entirely new breeds, while others simply wanted to enhance existing ones.

In the 1920s, for instance, the Argentinian Dr. Antonio Nores Martinez wanted to create an excellent hunter who could serve as an effective family guardian. He cross-bred several breeds, including Bull Terriers, Mastiffs, Great Danes, and even the Cordoba Fighting Dog – which no longer exists – until he created the Dogo Argentino. This dog was banned in the UK due to its high levels of aggression.

The creation of the Doberman Pinscher, which remains one of the most effective protection dogs today, is another interesting story. The development of this breed all started when a German tax collector was desperate for a guardian animal to accompany him at work. He cross-bred Beaucerons, Greyhounds, Rottweilers, German Pinschers, and Weimaraners to come up with the perfect guard animal for his needs.

The Development of Police Dog Programs

As more dog breeds were recognized, the police and military recognized the potential within certain driven, highly protective, tenacious, and intelligent dog breeds. Police departments in different parts of the world started conducting dog training programs in the early 1900s. The most common breeds used in these programs were the Belgian, Dutch, and German Shepherd dogs.

Recognizing the Opportunity for Training Dogs

The incredible role that police and military dogs play garnered the attention of many people, probing the question of whether guardian dogs could be employed for similar purposes in the private sector. Over the last few decades, the idea of using a guardian dog or other protective animal has become more systemized.

People seek out different types of dogs for distinct purposes, such as family, property, or livestock protection. Dogs are among the most popular guardian animals because they are relatively easy to train. When working with dogs, you can reinforce desirable behavior and get them to abandon negative ones. Training and guidance, combined with the natural bond they'll build with you, your family, and your livestock, will allow them to perform their roles effectively. The breed's temperament also influences how fit they are for their intended purpose.

The idea of using other animals, such as llamas, alpacas, donkeys, and even ducks, to protect certain types of livestock from specific types of predators is also growing more popular today.

The Advantages of Using a Guardian Animal as a Predator Management Method

Guardian dogs and other animals like llamas, alpacas, and donkeys have been used for centuries to protect livestock and property from predators, trespassers, and other potential threats. Depending on the nature of the guardian animal and the livestock they protect, guardians, especially dog guardians, often view the livestock they protect as companions. They learn to protect them out of care, which is something that the livestock also recognizes and responds to.

Why Do Guardians Protect Livestock?

Other guardian animals (like alpacas, donkeys, and llamas) often don't care about the livestock they protect. They would rather protect the flock because they either despise a potential predator or perceive it as an unknown source of potential danger. Training a guardian animal requires you to socialize it with the livestock it will be protecting. The socialization process allows the guardian to familiarize itself with the flock or herd. This way, the livestock becomes acceptable to the guardian and is no longer perceived as potentially dangerous or uncomfortable. However, when an unknown predator approaches, the guardian animal will lean into their instincts and react accordingly.

Regardless of whether the guardian animal protects the livestock out of care or mere familiarity, it will get the job done. That said, understanding the differences between each species' mechanisms to protect livestock will allow you to choose the right animal for your needs. Using livestock guardians is one of the few non-lethal predator management methods, which is why it's becoming increasingly popular. It encourages species to coexist, crucial for sustaining a healthy ecosystem. Using livestock guardians also reduces the need for lethal and brutal predator control methods.

How Does It Work?

Using guardian animals like dogs is an effective control method because the animal signals their presence instead of killing predators. Signaling refers to using scent and vocal communication to mark an animal's territory. When potential predators like foxes and wild dogs recognize that the guardian inhabits a certain location, the predators will avoid going there so a conflict doesn't occur. If the predator decides to ignore the signal and head over to the marked territory anyway, the guardian animal will show signs of aggression and attack the predator in case it persists. The livestock knows that if it flees, the predator will perceive it as a more attractive target, which is why it simply gathers behind its guardian, allowing it to do its job.

When employed correctly, guardian animals can prove to be very helpful at protecting predators, as well as animals like kangaroos and deer. While they successfully get rid of predators without killing them, they don't push them into nearby lands. Predators often free the marked territory altogether, which creates a safe space for the livestock to graze and wander around.

Things to Keep in Mind

When considering getting a dog or another guardian animal to protect your livestock, many worry that the guardian might harass or even harm the flock. This, however, isn't true. This predator control method is effective when landowners ensure that there are enough guardian animals for the size of the property and the number of livestock. They must also consider the nature of the terrain, the type of vegetation, the predator species and numbers, and the type of terrain and livestock.

Remember that your guardian animal is a living entity, not a robot. Just as all employees need healthy work environments and appropriate compensation, your animal needs to be rewarded for its efforts. It needs to be trained and guided so it can thrive at its job. You should neither treat the animal as a house pet nor neglect it altogether. You must ensure the guardian animal's physical and psychological needs are met. Providing the guardian animal with proper care ensures its health and longevity and allows it to do its job effectively.

Remember that, at first, you'll have to invest a lot of time, effort, and money into your guardian animal. Your investments, however, will pay back over time with less predator threats and more peace of mind. If you decide that you no longer need the guardian animal, you can also sell it for profit later.

When choosing a guardian animal, you should know that each has *its own limitations.* Some animals require more training and maintenance than others. Others require more support and positive reinforcement. If you need a quick fix to your problem, you shouldn't opt for a dog because they require a lot of training and support. For instance, if you're searching for a low-maintenance guardian animal, then a guardian donkey wouldn't be your safest bet.

The rewards you can reap from keeping a guardian animal don't come at a small price. This is why, when exploring your options, you should consider the amount of initial time, effort, and financial investment you're willing to make and the level of maintenance and support you're willing to provide the animal over time.

Ethical and Legal Considerations

If you plan on owning a guard animal, remember that the animal might be considered dangerous by law, depending on your location and the type of animal. For instance, guard dogs, even after retirement, are considered dangerous under the Domestic Animals Act of 1994. You must comply with certain regulations and conditions when keeping a guard animal; otherwise, you'll likely be subject to penalties for non-compliance.

Note that the following regulations can vary depending on where you live, so check with your local authorities before acquiring a guard dog. However, the following are the most common rules and non-compliance penalties:

- If your guardian dog attacks someone, you can get jailed for 5 to 10 years under the Crimes Act of 1958.
- You must notify the authorities about the location of your guard animal.
- You must follow the applicable enclosure requirements.
- You should always keep the animal collared, on a lead, and muzzled when accompanied outside the property.
- If you plan on selling or putting the animal up for adoption, prospective buyers or adopters must be notified that the animal is dangerous. This must be stated in writing as well.

- If the animal owner is under 18 years old, the owner's guardian will be considered the legal owner of the animal.
- Some guard animals must be implanted with an ISO microchip in accordance with the applicable regulations.
- The responsible authorities must be provided with the chip's ID number, alongside other required information.
- Guard animals must always wear a prescribed, reflective collar that can be recognized from a distance. The collar is usually red and yellow and between 20 and 30 mm wide.
- You must put up signs alerting visitors and passers-by that you're keeping a dangerous animal on your property. The signs must be clear, durable, reflective, and placed on all property entrances.
- If your dog is not guarding a non-residential property, it must be kept in a secure enclosure that prevents it from escaping.
- Only the legal owners of the animal must have access to the enclosure.
- The enclosure must be spacious, granting the animal enough room. It must also be supplied with a roof, a drain, a weatherproof sleeping area, walls, and a gate with a lock.
- Check with your local authorities about the details and requirements for the enclosure construction. The usual requirements for a dangerous dog's fence, however, are that it must be at least 2.63 feet tall and made of timber, brick, iron, concrete, or any other strong material mixed with chain mesh. You should regularly maintain the enclosure to ensure it's not weakened or damaged over time, compromising your dog's safety or granting them a way to escape. All the gates must be locked while the dog is on guard.
- You must notify the local authorities if your dangerous pet goes missing, gets adopted or sold, or its owner changes for any reason, or if you move into another home or relocate the pet. *In most cases, you must inform the relevant entity within 24 hours of these changes.*

After reading this chapter, you should better understand what it means to have a guard animal protecting your property or livestock. Guard animals can be a great, non-lethal method of predator control when utilized correctly. The effectiveness of this technique depends on your choice of animal and how well it suits your personal goals and resources, along with how it bonds with and protects your livestock and the type and number of predators in the region. There are also some ethical and legal concerns that you must keep in mind when getting a guard animal. This way, you'll avoid getting into trouble and ensure your animal is psychologically and physically prepared to fulfill its role effectively.

Chapter 2: Predator Behavior and Control Strategies

In these intricate ecosystems, predator behavior plays a pivotal role in shaping the dynamics of natural communities. Predators are an integral ecosystem component, exhibiting a fascinating array of behaviors honed by evolutionary processes to ensure their survival and success as hunters. From intricate hunting strategies to establishing territories, predator behavior is a complex interplay of instinct, adaptation, and environmental influence. This chapter explores the fundamental aspects of predator behavior, seeking to uncover the mysteries of how these creatures navigate their environments, select and pursue prey, and establish their roles within the delicate balance of ecosystems.

Fox preying on a mole.
Lauris Rubenis, CC BY 2.0 <https://creativecommons.org/licenses/by/2.0>, via Wikimedia Commons. https://commons.wikimedia.org/wiki/File:Fox_eating_mole.jpg

Common Predators to Know

Arctic Fox (Vulpes lagopus)

Arctic foxes have thick fur that changes color with the seasons, helping them blend into their snowy surroundings. They have a compact build and a bushy tail for balance. Furthermore, these animals are highly adapted to cold environments. Arctic foxes are known for their intelligence and resourcefulness. They are opportunistic omnivores, scavenging for food when necessary – and primarily feeding on small mammals, birds, and insects. They use their keen sense of hearing to locate prey beneath the snow, then pounce to catch it. These foxes can be expected visitors if you live in areas with heavy snowfall.

Bengal Tiger (Panthera tigris tigris)

Bengal tigers are large and muscular, with a distinctive orange coat marked by black stripes. They have powerful jaws and retractable claws. Solitary by nature, Bengal tigers are territorial and can be nocturnal and diurnal. They are strong swimmers and climbers. Tigers are ambush predators, relying on stealth and strength to get close to their prey before launching a powerful attack. Their preferred prey includes deer, wild boar, and sometimes larger animals like buffalo. These tigers can feed on livestock being raised near their habitat.

Golden Eagle

Golden eagles are large raptors with distinctive white heads, tails, a pointy beak, and sharp talons. They have powerful wings for soaring. Known for their keen eyesight, bald eagles are opportunistic hunters and scavengers. They are often found near large bodies of water. Bald eagles primarily hunt fish, using their powerful talons to snatch prey from the water. They also feed on livestock, waterfowl, and scavenged carrion. You can always expect bald eagles and other flying carnivores to attack poultry, rabbits, and other smaller animals they can easily hunt.

Cheetah (Acinonyx jubatus)

Cheetahs are slender and built for speed, with distinctive black tear stripes on their face. They have a lightweight frame, long legs, and a flexible spine. They are the fastest land animals, capable of reaching up to 75 miles per hour. They are primarily solitary, with males forming small groups called coalitions. Cheetahs rely on incredible speed and agility to chase down prey, mainly small to medium-sized ungulates like gazelles. Their hunting strategy involves stalking and a sudden acceleration to catch their prey. They are a potential predator of animals like sheep, goats, and cattle raised near their habitat.

Komodo Dragon (Varanus komodoensis)

The Komodo dragon is the world's largest lizard, with a robust build, scaly skin, and a long, muscular tail. Their saliva contains bacteria, making their bites potentially deadly. Komodo dragons are solitary and territorial. They are skilled climbers and swimmers despite their large size. These opportunistic predators prey on various animals, including deer, birds, and smaller dragons. If livestock is present within their territory, a Komodo dragon won't hesitate to make your livestock their next meal. They use their powerful jaws and sharp teeth to deliver a lethal bite, and their venomous saliva aids in incapacitating prey over time.

Snow Leopard (Panthera uncia)

Snow leopards are elusive and solitary.
Bernard Landgraf, CC BY-SA 3.0 <http://creativecommons.org/licenses/by-sa/3.0/>, via Wikimedia Commons: https://commons.wikimedia.org/wiki/File:Uncia_uncia.jpg

Snow leopards are adapted to cold, mountainous environments. They have a thick, fur-covered tail for balance, large paws for walking on snow, and distinctive rosette markings on their fur. Snow leopards are elusive and solitary, well adapted to the harsh conditions of high-altitude regions. They are known for their excellent leaping and climbing abilities. These animals primarily prey on blue sheep and other mountain ungulates. They use their powerful hind limbs to make incredible leaps and ambush their prey from above. If you raise livestock in mountainous regions with lots of snowfall, you can expect snow leopards if their habitat is nearby.

Red Fox (Vulpes vulpes)

Red foxes have a slender body, a bushy tail, and distinctive red fur. They are adaptable, found in various habitats, and are referred to as opportunistic omnivores. They are known for their intelligence and ability to thrive in both urban and rural environments. Red foxes have a diverse diet, including small mammals, birds, insects, and fruits. They use their keen senses, agility, and stalk-and-pounce hunting techniques to catch prey.

Spotted Hyena (Crocuta crocuta)

Spotted hyenas have a robust build, powerful jaws, and a distinctive hunched back. They have a unique social structure with a matriarchal hierarchy. They are highly social and live in clans. They are known for their vocalization and often compete with other predators for food. Spotted hyenas are skilled scavengers and hunters, preying on various animals, including wildebeests and zebras. They use cooperative hunting techniques to exhaust and bring down larger prey.

Various other predators can potentially feed on domestic and livestock animals. These include snakes, coyotes, skunks, bobcats, raccoons, and hawks.

Predator Identification

Predator identification is crucial for implementing effective protection measures, as it allows for targeted conservation efforts and the mitigation of potential conflicts. Here are some general guidelines to enhance predator identification.

In-Depth Study of Field Guides

Acquire and thoroughly study field guides specific to the geographical region of interest. Pay attention to detailed descriptions, illustrations, and range maps for each predator species. Understand variations in physical characteristics based on age, sex, and season.

Comprehensive Training and Workshops

Attend specialized training sessions or workshops conducted by wildlife experts. Participate in hands-on activities, including specimen examination and practical field exercises. Engage in discussions on the nuances of predator identification, emphasizing distinguishing features.

Online Resources Exploration

Explore reputable online resources provided by wildlife organizations and research institutions. Utilize multimedia content, including videos, images, and interactive tools. Familiarize yourself with digital databases and citizen science projects contributing to species identification efforts.

Engagement with Local Experts

Establish connections with local wildlife experts, researchers, and naturalists. Arrange field visits or guided tours with these experts to gain real-world exposure to predator habitats. Seek mentorship to refine your observational skills and receive personalized guidance.

Mastery of Tracks and Signs

Develop a deep understanding of tracks, scat, and other signs left by predators. Engage in tracking courses to enhance your ability to interpret these signs accurately. Practice distinguishing between similar-looking tracks and recognizing patterns that indicate specific species.

Behavioral Analysis

Study the behavioral ecology of different predators. Understand nuances in hunting techniques, communication methods, and social structures. Observe captive animals to witness behaviors that might be challenging to observe in the wild.

Technological Integration

Embrace technology, including the use of camera traps and trail cameras. Learn how to set up and maintain these devices effectively. Familiarize yourself with image recognition software and understand its limitations and strengths in predator identification.

Identification of Key Indicator Species

Identify and deeply understand key indicator species relevant to the ecosystem. Explore their ecological roles, habitat preferences, and interactions with other species. Use these indicators to gauge ecosystem health and identify potential threats to predators.

Systematic Documentation

Develop a systematic approach to documenting predator observations. Include detailed notes on behavior, location, and environmental conditions. Capture high-quality photographs and record audio if possible. Share your observations through scientific channels, online platforms, or local conservation initiatives.

Predator Control

Predator control refers to managing and mitigating predators, typically through human intervention, to minimize their impact on livestock and property. This practice is essential for safeguarding agricultural interests, protecting domestic animals, and maintaining ecological balance. The primary goals of predator control are to reduce economic losses, ensure the safety of domestic animals, and prevent potential conflicts between wildlife and human activities. Several reasons underscore the importance of predator control.

Livestock Protection

One of the primary reasons for predator control is to safeguard livestock from predation. Predators like wolves, coyotes, big cats, and bears may threaten domestic animals like sheep, cattle, and poultry. Reducing the impact of predators helps maintain the livelihoods of farmers and ranchers who depend on livestock for income.

Economic Considerations

Predation of livestock can result in significant economic losses for farmers and ranchers. Besides the direct loss of animals, predation can lead to decreased productivity, lower reproductive rates, mental health issues in farm animals, and increased costs associated with additional security measures. Effective predator control minimizes these economic impacts.

Human Safety and Property Protection

Certain predators, especially those living near human habitats, can threaten human safety and property. For example, large predators may enter residential areas searching for food, potentially leading to conflicts.

Predator control measures help mitigate these risks and promote coexistence between humans and wildlife.

Conservation of Endangered Species

Predator control is not solely about eradicating predators; it often involves carefully managing their populations. This is crucial for conserving endangered or threatened species by addressing conflicts with human activities. Balancing predator populations with the needs of local communities contributes to overall biodiversity conservation, eradicating the population from a habitat that would otherwise disrupt the food chain and the ecosystem.

Ecological Balance

Uncontrolled predator populations can disrupt ecological balance. In some cases, overpopulation of certain predators can lead to declines in prey species, affecting vegetation and other ecosystem components. When implemented responsibly, predator control helps maintain a healthier balance within ecosystems.

Livestock Industry Sustainability

The sustainability of the livestock industry is closely tied to effective predator control. By mitigating predation risks, farmers and ranchers can maintain stable and profitable operations, contributing to the sustainability of the agriculture sector.

Ethical Considerations

While predator control is sometimes necessary for practical reasons, ethical considerations are essential. Using humane and ethical methods to minimize harm to predators and non-target species is necessary. Balancing the needs of human livelihoods with the ethical treatment of wildlife is the key to effective and sustainable predator control.

Predator Control Plan

Creating a predator control plan requires a thorough understanding of the specific circumstances, including the types of predators involved, the nature of the environment, and the community's or stakeholders' goals. Here's a step-by-step guide with corresponding examples to help tailor a predator control plan to specific circumstances:

Step 1: Assess the Situation

In a region where coyotes threaten livestock, assess the specific risks by considering the size of the coyote population, the types of livestock present, and historical predation patterns.

Step 2: Identify Target Predators

Determine whether the main concern is large predators like wolves, smaller predators like foxes, or a combination of both. Tailor the control plan based on the specific predators causing issues.

Step 3: Understand Local Ecosystem Dynamics

In an area with a delicate ecosystem, consider the potential impacts of predator control on non-target species and ecological balance. Assess how changes in predator populations may affect prey species and vegetation.

Step 4: Set Clear Goals and Objectives

Define specific goals, such as reducing livestock losses by a certain percentage or promoting coexistence between predators and livestock. Clear objectives help measure the success of the control plan.

Step 5: Choose Appropriate Control Methods

Depending on the circumstances, control methods could include non-lethal measures like improved fencing, guard animals, or deterrents. If lethal measures are deemed necessary, choose methods that are humane and targeted, such as selective trapping or shooting in specific situations.

Step 6: Consider Non-Lethal Alternatives

Explore using deterrents like sound or light devices, guardian animals (e.g., llamas, dogs), or modifying livestock management practices (e.g., changing grazing patterns) to reduce predation without harming predators.

Step 7: Implement Monitoring and Evaluation

Establish a system to monitor the effectiveness of control measures and adjust the plan accordingly. This might involve regular surveys of predator populations, tracking livestock losses, and assessing changes in ecological indicators.

Step 8: Involve Stakeholders and the Community

Engage with local farmers, ranchers, conservationists, and other stakeholders to gather insights and address concerns. Community involvement fosters a collaborative approach and increases the plan's chances of success.

Step 9: Educate and Raise Awareness

Conduct workshops, training sessions, or informational campaigns to educate the community about predator behavior, the importance of coexistence, and the rationale behind specific control measures.

Step 10: Adaptability and Flexibility

Be prepared to adjust the control plan based on evolving circumstances. For instance, if a new, non-lethal technology becomes available, consider integrating it into the plan to improve effectiveness.

Step 11: Legal and Ethical Considerations

Ensure that the control plan complies with local and national regulations. Consider the ethical implications of control methods and strive to use approaches that minimize harm to predators and non-target species.

Step 12: Documentation and Reporting

Keep detailed records of control activities, outcomes, and any unexpected consequences. Regularly report findings to stakeholders, funding agencies, and regulatory bodies to maintain transparency.

Following these steps and tailoring each aspect to the specific circumstances can enable you to make a predator control plan and effectively address the challenges while considering the ethical implications of predator management. The key is to strike a balance that protects human interests and wildlife conservation.

Non-Lethal Predator Control Strategies

Improved Fencing

Reinforce existing fencing or install predator-specific barriers to prevent access to livestock. It's a humane method as it keeps predators away without harming them. Installing these fences won't be an issue for local authorities and is a minimal method that aids habitat preservation.

Guardian Animals

Introduce trained animals like dogs, llamas, or donkeys to deter predators from approaching livestock. Your livestock won't be bothered as these animals have natural deterrent instincts to stop predators from coming nearby. Getting trained dogs for protection won't be an issue. This positive approach to predator control promotes coexistence without direct harm to predators.

Deterrent Devices

- Use devices emitting sounds, lights, or odors to discourage predators from entering an area. You won't be harming predators with these gadgets while ensuring the domestic animals stay protected. However, before using any of these devices, please contact the local authorities, as there are restrictions for using specific deterrent devices. Although this method effectively keeps predators at bay, these devices can also create disturbances for other wildlife.

Livestock Management Practices

Implement strategies like changing grazing patterns or providing night penning to reduce predation risk. Executing this method will be possible through a change in livestock management practices where you reduce the exposure of the prey to the predators.

Lethal Predator Control Strategies

Selective Trapping

Use traps to capture specific predators, allowing for their release or euthanasia. There are mixed opinions about trapping, and regulations must be followed in most regions.

Culling

This is the controlled reduction of predator populations through targeted removal. It's a controversial practice and must be conducted ethically with the involvement of relevant authorities. If the predator is a wildlife species, go through the wildlife management laws and the rules you must comply with. Eradicating predator populations in a specific area can potentially disrupt the balance of the ecosystem.

Aerial Shooting

Employ aircraft to shoot predators from the air. This method is mainly used for large predators when attempts for ground capture fail. A dart gun is used to inject a sedative into the animal, making them unconscious and enabling their safe capture. Before using this method, you will need a special permit or a heads-up from the local authorities.

Toxicants/Poisons

You can also use chemical substances to poison predators. This method is considered inhumane and controversial, as it kills the predators in an inhumane manner, puts other animals at a higher risk of secondary poisoning, and can significantly disrupt the ecosystem.

Monitoring and Adaptive Management

Regularly assess the effectiveness of control strategies and adjust them based on results. With regular monitoring, you'll naturally minimize potential risk and improve your existing strategies if any discrepancy comes up.

Public Education and Outreach

Inform and engage the public about the rationale and methods of predator control. You can educate your neighbors and community members about the methods of predator control, establish their necessity, and determine the positive effects they will have on every species involved when done the right way.

Implementing a combination of these strategies while carefully considering ethical, legal, and environmental factors can contribute to effective and responsible predator control while minimizing negative impacts on ecosystems and wildlife. Regular assessment, community involvement, and adherence to ethical standards are essential for the success of any predator control program.

Integrated Predator Management

Integrated Predator Management (IPM) is a holistic approach that combines a variety of strategies to manage and mitigate the impact of predators on livestock, wildlife, and ecosystems. The key principle of IPM is to use a combination of non-lethal and lethal methods in an integrated and adaptive manner, considering the ecological, economic, and ethical aspects of predator control. This approach aims to achieve maximum effectiveness in reducing predation while minimizing harm to non-target species and maintaining a balance in the ecosystem.

Real-Life Examples of Successful Integrated Predator Management

Yellowstone National Park, USA

The reintroduction of gray wolves into Yellowstone in 1995 led to a shift in the ecosystem. To address conflicts with local ranchers, a combination of range riders, guardian animals (such as dogs and llamas), and targeted lethal control was employed. Integrating non-lethal and lethal measures contributed to reduced livestock predation and facilitated the coexistence of wolves and ranching communities.

NamibRand Nature Reserve, Namibia

In NamibRand, where cheetahs were preying on livestock, a combination of guard dogs, improved fencing, and community engagement initiatives were implemented. Non-lethal measures significantly reduced livestock losses, fostering a more positive relationship between the local community and the cheetah population.

France – Pyrenees Mountains

In regions where brown bears were preying on sheep, an integrated approach included guardian dogs, electric fencing, and compensation programs for affected farmers. Outcome: The combined measures led to a reduction in bear-caused livestock losses, contributing to both predator conservation and the protection of livestock.

Australia – Livestock Guardian Dogs

Livestock guardian dogs, such as the Kangal and Maremma breeds, are employed to protect livestock from foxes and dingoes. These dogs have successfully reduced predation on sheep and other livestock without resorting to lethal control, demonstrating the effectiveness of non-lethal methods.

IPM recognizes the importance of maintaining ecological balance by considering the broader impacts of control measures on non-target species and ecosystem health.

With its multifaceted and adaptive approach, Integrated Predator Management demonstrates that effective predator control can be achieved while promoting coexistence and minimizing negative impacts on predators and ecosystems. Success stories from various regions emphasize the importance of tailoring strategies to local circumstances and involving communities for sustainable and responsible outcomes.

Chapter 3: Choosing the Right Guard Animal

A comprehensive assessment of protection needs is a pivotal starting point in safeguarding livestock and property. This foundational chapter aims to enlighten you on the significance of this evaluation, emphasizing its role in informed decision-making when selecting guard animals and crafting protective strategies.

Before exploring guard animals and protective strategies, it's imperative to recognize the essence of conducting a thorough assessment. This process serves as the foundation upon which adequate protection is built. It involves tapping into factors like the type of livestock, geographical location, prevalent threats, and specific vulnerabilities unique to the property. Protection isn't a one-size-fits-all endeavor. Each livestock species demands a tailored approach. It starts with an assessment that includes understanding these threats, whether predatory wildlife, opportunistic thieves, or environmental challenges is paramount. Geography also plays a pivotal role in determining protection needs. The challenges faced by livestock and property in different regions vary significantly.

This chapter lays the groundwork, developing an understanding of the critical role played by a thorough assessment in devising effective protection plans. Armed with this foundational knowledge, you will be better equipped to navigate the nuanced world of livestock and property protection as you progress through the book.

Understanding Livestock Dynamics

To comprehensively understand the protection needs of your livestock, start by dissecting each species' behaviors, habits, and specific requirements. Cattle, for example, operate within a social hierarchy and may require protective measures that consider their herd dynamics. Sheep and goats, known for their flock mentality, might need different strategies. Poultry, with diverse breeds and foraging habits, demands a nuanced approach. Delve into the intricacies of their daily routines, feeding habits, and interactions to tailor protection that aligns seamlessly with their unique needs.

Cattle operate within a social hierarchy and may require protective measures.

https://unsplash.com/photos/herd-of-cattle-on-grass-field-during-daytime-0vVQWN_D26c

Property Dimensions

Beyond livestock specifics, the physical layout of your property plays a critical role in designing adequate protection. Evaluate the size of your property, noting features like natural barriers, open spaces, and secluded areas. Identify potential hiding spots for predators or areas where livestock may congregate. This scrutiny enables the strategic placement of guards and aids in designing fencing that complements the existing layout. An in-depth assessment of these physical dimensions provides the groundwork for a protection plan harmonizing with the terrain.

Geographical Insights

Geography forms the canvas upon which your protection strategies are painted. Assess the terrain – whether your property spans hilly landscapes, flat plains, or includes water features. Consider the climate, understanding temperature ranges, precipitation levels, and seasonal variations. Factor in regional nuances, such as nearby wildlife or prevalent environmental challenges. This comprehensive understanding guides decisions on guarding animal selection and implementing protective infrastructure that is seamlessly woven into the natural environment.

Microclimate Considerations

Identifying microclimates adds another layer of precision to your protection plan. Identify areas on your property that may experience unique weather patterns or temperature variations. Recognize spots that may be more prone to extreme conditions. Tailoring protection to these microclimates ensures that your strategies are practical and adaptable to the specific nuances of your property's climate. This granular approach enhances the well-being of your livestock by acknowledging and accommodating localized variations.

Documenting Vulnerabilities

Analyze your property critically to identify vulnerabilities in your existing protection measures. Look for potential weak points in fencing, areas prone to erosion, or access points for predators. Documenting these vulnerabilities serves as a roadmap for strategic fortification. This proactive approach minimizes risks, whether it involves reinforcing fences, implementing deterrents, or redesigning specific areas. It strengthens the overall robustness of your protection plan.

Evaluating Your Existing Protection Measures

As you safeguard your livestock and property, rigorous evaluation of your existing protection measures is imperative. This reflective process ensures that your current strategies are practical and aligned with your environment's evolving needs and challenges. Here's a step-by-step guide to aid you in this critical self-assessment:

Fencing Infrastructure

Begin by closely examining your fencing infrastructure. Assess its integrity, height, and material. Identify any signs of wear, damage, or potential weak points. Evaluate whether the fencing aligns with the specific needs of your livestock and property layout. Consider the type of predators in your region and assess whether your current fencing is a sufficient deterrent. Conduct periodic checks to ensure ongoing effectiveness.

Lighting Systems

Evaluate the lighting systems implemented on your property, especially in areas crucial for livestock and property protection. Adequate lighting not only deters potential threats but also aids in surveillance. Assess the coverage, brightness, and reliability of your current lighting installations. Consider whether they effectively illuminate vital areas and potential blind spots. Optimal lighting enhances overall security and can be a valuable deterrent during nighttime hours.

Alarm Systems

If you have alarm systems in place, scrutinize their functionality and coverage. Test the alarms periodically to ensure they are in working order. Assess whether the alarm systems cover critical areas and are equipped to detect potential threats promptly. Evaluate the responsiveness of your monitoring or response mechanisms tied to these alarms. A robust alarm system serves as an early warning mechanism, providing a crucial layer of protection.

Surveillance and Monitoring

Consider any surveillance or monitoring systems integrated into your protection measures. This could include security cameras, motion sensors, or other technological aids. Evaluate the coverage and clarity of your surveillance systems. Ensure that monitoring is consistent and that potential blind spots are addressed. Embrace technological advancements that align with your protection needs, enhancing your ability to detect and respond to threats effectively.

Integration of Guard Animals

For those incorporating guard animals into their protection plan, evaluate the effectiveness of their presence. Assess whether the selected species align with the specific threats in your region. Observe their behavior and interactions with livestock. Gauge whether they contribute to a sense of security and act as a deterrent to potential predators. The symbiotic relationship between guard animals and other protective measures is pivotal for a holistic defense strategy.

Response Protocols

Evaluate the response protocols that are in place. Assess whether you have clear and well-communicated action plans in case of a security breach. Review the coordination among individuals involved in responding to threats. Ensure that everyone on your property knows their roles and responsibilities during such situations. Regularly conduct drills to practice response protocols and address any areas needing improvement.

Regular Maintenance

Lastly, consider the maintenance aspect of your protection measures, regularly scheduled inspections, and maintenance checks for fencing, lighting, alarm systems, and other components. Address wear and tear promptly and replace or upgrade equipment as needed. A proactive approach to maintenance ensures that your protection measures remain reliable and effective over time.

Noise and Disturbance Analysis

Consider the potential impact of noise and disturbance on the functionality of your protection measures. Noise, whether from nearby construction, machinery, or natural elements like storms, can compromise the effectiveness of alarms or the vigilance of guard animals. Assess the decibel levels and frequency of these disturbances and strategize ways to mitigate their impact on your security systems.

Vegetation Management

Evaluate the role of vegetation in your protection infrastructure. Overgrown bushes or trees can compromise the visibility of security cameras and provide hiding spots for potential threats. Regularly assess and trim vegetation to maintain clear sightlines and eliminate potential blind spots, ensuring that natural growth does not hinder your protective measures.

Multi-Layered Approach

Assess how well your protection measures work together in a synergistic, multi-layered approach. Ensure that the various elements – fencing, lighting, alarms, and guard animals – complement each other to create a robust defense system. Consider potential gaps or overlaps in coverage and fine-tune your strategy to achieve optimal synergy among the different layers of protection.

Neighbor Collaboration

Explore collaboration with neighboring properties to enhance collective security. Evaluate the possibility of sharing a communication network or coordinating protection efforts. Open communication with neighbors can foster a sense of community protection, enabling a quicker and more effective response to potential threats that transcend property boundaries.

Adaptability to Emerging Threats

Anticipate and assess how well your protection measures can adapt to emerging threats. Stay informed about new predator species in the region, evolving environmental conditions, and technological advancements that could impact your defense strategies. Develop a proactive mindset that allows your protection plan's continual evolution to address emerging challenges effectively.

Accessibility for Emergency Services

Evaluate the accessibility of your property for emergency services. Ensure that entrances are marked and paths are easily navigable. This consideration is crucial for facilitating a swift response from emergency services in case of incidents requiring their intervention, thus contributing to your livestock's overall safety and well-being.

Livestock Identification Measures

Assess the adequacy of your livestock identification measures. This includes ear tags, microchips, or other identification methods. Clear and compelling identification is essential for managing your livestock. It plays a crucial role in recovery efforts in the unfortunate event of theft or loss.

Wildlife-Friendly Measures

Evaluate whether your protection strategies consider the well-being of local wildlife. Implement measures that deter threats to livestock while minimizing harm to native species. This could involve using wildlife-friendly fencing or deterrents that specifically target potential predators without causing harm to non-threatening wildlife.

Cybersecurity Measures

If you have technologically advanced protection systems, assess their vulnerability to cyber threats. Implement robust cybersecurity measures to safeguard against unauthorized access or tampering. Regularly update software and firmware to address potential vulnerabilities and stay ahead of evolving cyber threats, as digital systems used for livestock management and security systems can be a potential target.

Legal Compliance

Ensure that your protection measures comply with local regulations and zoning laws. This includes adhering to any restrictions on fencing height, lighting, or the use of guard animals. Legal compliance avoids potential legal issues and ensures your protective measures align with community standards and regulations.

By meticulously exploring these facets in extreme detail, you gather information and develop a comprehensive understanding, which is the bedrock for informed and tailored protection strategies. This depth of insight positions you to navigate the complexities of livestock and property protection with precision and adaptability.

Clarifying Specific Protection Needs

In pursuing a robust defense strategy, the clarity of specific protection needs becomes paramount. Each step in this process requires a nuanced understanding, allowing for the precise tailoring of defenses to mitigate identified risks. Let's delve into the details:

Prioritize Critical Areas

Initiate by meticulously identifying and prioritizing critical areas on your property. These could range from vulnerable entry points to zones with high livestock concentration or areas susceptible to environmental risks. This strategic prioritization forms the foundation for targeted resource allocation. For prioritized critical areas, elevate the fortification of fencing and physical barriers. Choose materials and designs that align precisely with identified threats. Heighten the structure to a level that acts as an effective deterrent against predators.

Evaluate Livestock Concentration

Identify areas where livestock concentration is pronounced. These spaces demand heightened protective measures. Consider strategies such as reinforced fencing, increased lighting, or advanced surveillance to fortify these zones against potential threats.

Identify High-Risk Zones

Scrutinize the landscape to identify high-risk zones based on the previously assessed vulnerabilities. Dense vegetation, uneven terrain, or proximity to natural habitats could elevate risks. Tailoring defenses in these zones employs strategies like vegetation clearance and specialized fencing.

Customize Protection for Specific Livestock

Precision is critical when customizing protection for each type of livestock. Recognize the unique vulnerabilities and behaviors associated with different species. Craft tailored strategies that address the distinct challenges posed by each livestock category.

Employ Advanced Surveillance in High-Risk Zones

High-risk zones demand a sophisticated surveillance approach. Deploy advanced technologies like infrared cameras or sensors with an extended range. This technological investment enhances monitoring and provides early detection capabilities, especially in areas susceptible to environmental challenges.

Integrate Guard Animals Strategically

Strategically position guard animals based on the prioritized critical areas. Optimize their locations to cover zones effectively. Rotation of guard animals may be necessary to maximize their impact on deterrence, aligning with the ever-changing landscape of potential threats.

Monitor Environmental Changes

Maintain a vigilant stance toward environmental changes that may impact protection needs. Seasonal variations, migration patterns of potential threats, or alterations in vegetation should prompt adaptive adjustments to your defense strategy.

Invest in Targeted Technologies

Invest strategically in technologies specifically designed to counter identified risks. This may include specialized alarms, tracking devices for high-value livestock, or technologies tailored to mitigate environmental risks. Precision in technology investments ensures optimal resource utilization.

Evaluate Natural Barriers

Conduct a comprehensive assessment of the natural barriers present on your property, such as rivers, cliffs, or dense forests. Understand their potential as deterrents for unauthorized entry or predators. Simultaneously, evaluate how these features might create challenges or opportunities for your livestock as well as potential threats. Consider factors such as ease of access for wildlife or the impact of terrain on visibility.

Implement Biosecurity Measures

Integrate a robust biosecurity system to prevent the introduction and spread of diseases within your livestock. This involves implementing strict quarantine protocols for new animals, conducting regular health checks, and ensuring proper waste disposal practices. Evaluate the effectiveness of these measures in maintaining the health and well-being of your livestock while considering factors like disease prevalence in your region.

Enhance Communication Systems

Upgrade your communication systems to ensure swift and effective coordination during emergencies. This may involve investing in reliable two-way radios, mobile phones with consistent coverage, or community alert systems. Evaluate the reach and reliability of these systems while considering factors such as terrain and potential interference.

Utilize Smart Technologies

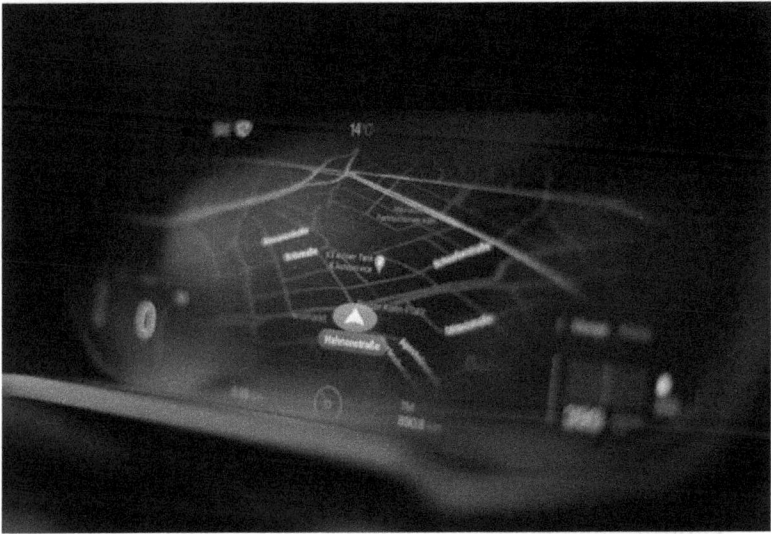

A GPS tracking device can help you monitor your livestock's movements.
https://unsplash.com/photos/black-and-white-car-dashboard-p79nyt2CUj4?utm_content=creditShareLink&utm_medium=referral&utm_source=unsplash

Explore the integration of smart technologies to enhance your overall security measures. Consider employing GPS tracking devices for your livestock to monitor their movements in real-time. Additionally, assess the feasibility of using drones for aerial surveillance, providing a bird's-eye view of your property. Evaluate the practicality and effectiveness of these technologies in your specific context.

Investigate Local Wildlife Patterns

Gain a deep understanding of your area's local wildlife patterns and behaviors. This knowledge is crucial for crafting defenses, considering interactions between domestic animals, wildlife, and potential predators. Evaluate how wildlife behaviors might impact your livestock and adapt your protective measures accordingly. Consider factors such as migration patterns, feeding habits, and potential conflicts.

Establish Secure Storage

Create secure storage facilities for valuable equipment, feed, and other essentials. Assess the design and construction of these storage areas, considering factors such as durability, accessibility, and resistance to tampering. Evaluate how these secure storage spaces contribute to property security and protection against theft or damage.

Engage in Continuous Training

Promote ongoing and comprehensive training programs for individuals involved in livestock management. This includes educating them on response protocols and providing continuous training on identifying potential risks and implementing preventive measures. Evaluate the effectiveness of training programs through regular assessments and feedback mechanisms.

Investigate Animal Theft Prevention Measures

If livestock theft is a concern, take specific measures to prevent such incidents. This could involve implementing identification methods, such as microchipping or unique markings, to make stolen animals easily traceable. Evaluate the success of these prevention measures by tracking incidents and recovery rates.

Develop Relationships with Authorities

Establish and nurture relationships with local law enforcement and animal control authorities. Evaluate the effectiveness of these relationships in creating a collaborative approach to addressing security concerns. Assess the support and insights authorities provide to enhance overall livestock and property security.

Monitor Weather-Related Risks

Consider and monitor weather-related risks that might affect your livestock and property. Develop strategies to safeguard against flooding or extreme weather events, including elevated platforms for shelter or flood-resistant enclosures. Evaluate the resilience of these strategies in

---

the face of varying weather conditions.

Invest in Training for Guard Animals

Provide specialized and ongoing training for guard animals to enhance their effectiveness as protectors. This could include obedience training, exposure to various environmental conditions, and reinforcement of desired protective behaviors. Evaluate the impact of training on the behavior and performance of guard animals through regular assessments.

Explore Insurance Options

Investigate insurance options that cover losses related to livestock and property security. Evaluate the terms and coverage provided by different insurance plans, considering factors such as the financial impact of potential security incidents. Assess the reliability and responsiveness of insurance providers in the aftermath of security-related losses.

Understanding these factors enables you to thoroughly understand the nuances and intricacies of implementing these protection measures. This detailed evaluation forms the basis for a comprehensive and adaptive defense strategy tailored to your unique context and requirements.

Chapter 4: Livestock Guardian Dogs: Breeds and Traits

This chapter introduces you to livestock guardian dogs, exploring the different breeds and the traits that make them effective in protecting livestock and property. It explains why these animals are great livestock guardians, why breed selection is crucial when choosing livestock guardian dogs, and how certain breeds are better suited to specific environments and livestock. The chapter also provides a list of the most well-known guard breeds used as livestock guardians (along with their characters and temperament) and tips on choosing the right breed based on specific needs, climate, livestock, and more.

Dog as guardian.

The Importance of Livestock Guardian Dogs

Using guardian dogs to safeguard livestock is the most effective way of keeping predators like carnivores (both four-legged and winged) or other dogs at bay. In practice, dogs are typically used for guarding smaller animals, like poultry, and ruminants, like goats and sheep, as these are more vulnerable to predators due to their size. However, some dog breeds can also work with large-bodied livestock, like horses and cows.

Livestock guardian dogs have a few characteristics in common, including living and moving around the herd or flock. They're also larger in size and easy to work with, meaning they get along with the animals they are guarding, their owners, and other guard dogs, if there are any (large farms typically necessitate more than one livestock guardian dog). They have been bred to protect and avoid hurting the animals they guard, even if they're hungry or being attacked by them. At the same time, they're usually unfriendly with other animals, including stray and kept dogs that wander into their territory. A good guardian dog keeps even the neighbor's dogs and cats away, which is particularly good for protecting poultry. They also watch over the livestock's behavior, which can alert them of the presence of predators and other threats.

Even if you have unwanted visitors that won't harm your farm animals, they can still damage your property and steal food from your animals. Livestock guardian dogs will prevent this from occurring. They are independent thinkers, meaning they won't look for commands but take the action they are trained to take immediately upon noticing a potential threat.

Other benefits of using dogs as livestock guardians include an increase in profit (with fewer losses due to predators, animal husbandry becomes more profitable), proactive protection instead of mitigating damages, elimination of the need for other, more costly control measures, like traps and hunting, reducing the likelihood of the livestock coming in contact with wild animals carrying illnesses, and longer effectiveness at lower investment in labor and time.

Livestock guardian dogs are particularly effective with larger flocks and herds (keeping them for small groups may not be as cost-effective) in areas with high predation rates. When trained and cared for properly, canine guardians represent an excellent investment, especially for safeguarding small animals.

The Most Popular Livestock Guardian Dog Breeds

Not all dogs can become livestock guardians. Different breeds possess predator-specific traits that make them more effective at deterring particular types of predators. Below, you'll find the most popular livestock guardian dog breeds, their core characteristics, and information about socialization, livestock compatibility, and health.

Great Pyrenees

The Great Pyrenees is one of the largest dog breeds.

HeartSpoon, CC BY-SA 4.0 <https://creativecommons.org/licenses/by-sa/4.0>, via Wikimedia Commons: https://commons.wikimedia.org/wiki/File:Great_Pyrenees_Mountain_Dog_2.png

One of the largest guard dog breeds is the Great Pyrenees, which originates from the French and Spanish royal courts. Due to its calm and regal demeanor, this canine is most commonly used for protecting sheep. However, because it's always looking for action, it can work well with smaller animals, too – the more it has to walk around following them, the better. At the same time, Great Pyrenees are incredibly gentle and patient dogs, so you won't have to worry about them trampling poultry and young livestock. They're also perfect for family-owned farms as they are great with children, too.

Tall and built like a miniature horse, this dog has a long, double-layered, white fluffy coat and double dew claws on its rear limbs. They thrive in colder climates and can't stand higher temperatures because of their fine, wooly undercoat. They are also prone to hip dysplasia (abnormal tissue growth), especially if one or both of their parents had the condition.

Anatolian Shepherd

Anatolian Shepherds are known for their intelligence and loyalty.

Zeynel Cebeci, CC BY-SA 4.0 <https://creativecommons.org/licenses/by-sa/4.0>, via Wikimedia Commons: https://commons.wikimedia.org/wiki/File:Anatolian_Shepherd_Dog_01.jpg

Known for their intelligence and loyalty, Anatolian Shepherds are the embodiment of the perfect guard dog. They're naturally independent and calm but tend to bark excessively when they perceive a threat. This usually occurs when predators get too close; otherwise, these canines won't draw attention to themselves. If they do, you can be sure they're doing their job of protecting your animals and property. They can be paired up with livestock of all sizes, and you can be assured they'll keep them safe while you tend to your other responsibilities on your farm/house.

The Anatolians are large and muscular dogs with short fur, which means they fare well in hot weather. You can leave them alone with the herd or flock for several days, and they'll be fine. They have very few health issues, but they require intensive training. You have to be assertive when training them so they can learn what's required of them. Otherwise, they'll act independently, and although they're unlikely to hurt the livestock, they might damage your property in their overzealous attempt to chase away a perceived threat.

Komondor

Komondors are muscular canines.

Nikki68, CC BY 2.5 <https://creativecommons.org/licenses/by/2.5>, via Wikimedia Commons: https://commons.wikimedia.org/wiki/File:Komondor_delvin.jpg

Like the Anatolians, Komondors are muscular canines, which make them perfect for guarding large livestock or using them in areas where larger predators represent a frequent threat. Instead of barking, they launch into action and chase away predators or other intruders that wander onto your property. Their sheer size alone can be enough to deter small nuisances and for the livestock to respect them. They're very energetic, so they are recommended for larger territories where they can run around and actively safeguard larger herds or flocks of farm animals.

Komondors have a thick, white coat that is prone to matting. Still, it keeps them warm in the chilly winters in colder climates (they originate from Hungary). The unique appearance of their coat also helps them blend in well with sheep, which is why they are most commonly used for protecting these ruminants. Komondors require intensive training and socialization with the livestock before they can be trusted to safeguard them. Once the Komondor familiarizes itself with the herd or flock, they are tasked to protect, along with the territory, a strong bond is formed, and they'll go to great lengths to protect them. They are characterized by strong health and lengthy and active life.

Caucasian Shepherd Dog

Caucasian Shepherd Dogs were used to safeguard livestock.
Canarian, CC BY-SA 4.0 <https://creativecommons.org/licenses/by-sa/4.0>, via Wikimedia Commons: https://commons.wikimedia.org/wiki/File:CaucasianOwcha1.jpg

Hailing from Russia, the Caucasian Shepherd Dog was once used to safeguard livestock from large predators like wolves and bears. These are increasingly territorial canines with exceptional guard instincts. Once they perceive a threat, they'll take it on with confidence, so you won't have to worry about the safety of your livestock while busy with other tasks.

With their thick, double-layered fur, these dogs thrive in colder climates and aren't meant to be out in the sun all day in hot weather.

They have strong bodies, which makes them excellent protection against predators, but they're gentle enough not to harm small livestock. Still, in practice, they are typically used for guarding large or medium-bodied ruminants. They have few health issues and can live between 10 and 11 years.

Maremma Sheepdog

Maremma Sheepdogs have a muscular build.

Canarian, CC BY-SA 4.0 <https://creativecommons.org/licenses/by-sa/4.0>, via Wikimedia Commons: https://commons.wikimedia.org/wiki/File:Maremma_Sheepdog_male.jpg

The Maremma Sheepdog is a breed with rich, white fur and a muscular build meant for safeguarding livestock from predators of all sizes. While protecting the farm animals and bonding with them very easily, this canine won't hesitate to attack intruders. They are known to be particularly protective of young animals – and even remain wary of their owner's actions around youngsters. They're the perfect option for protection against four-legged predators like coyotes and foxes because they won't let themselves be outsmarted by these cunning creatures. The Maremmas are also suspicious and will sense a threat by observing the farm animals' behavior. They alert the intruders by barking loudly, ensuring they are heard and discouraging them from advancing into their territory.

They're perfect guards for sheep, especially if the animals are kept outside during mild and moderate winters. They also like to work in groups as they are very social, so they can be employed alongside other livestock guardian dogs. They have few health issues but are prone to hereditary hip dysplasia.

Kuvasz

The Kuvasz has Hungarian origins.
https://commons.wikimedia.org/wiki/File:Kuvasz_named_Kan.jpg

Like the Komondor, the Kuvasz also has Hungarian origins. They are fierce and strong canines, always ready to protect their territory and herd/flock. They form strong bonds with the owner and the livestock. While they won't hesitate to take action against a perceived threat, they aren't completely independent. They thrive with regular guidance and commands. At the same time, they are somewhat more challenging to train because they tend to establish a leadership position at an early age. For the same reason, they are best employed alone (without other guard dogs) or with canines they've been socialized with since puppyhood.

The Kuvasz are excellent choices for ranches and farms with large livestock breeds. They won't tolerate the aggressive behavior of other species, including farm animals like poultry, but they won't harm them inadvertently. They have a corded white coat specifically developed to blend in with goats and sheep that they have been used to guard since ancient times. They tolerate all temperatures fairly well and typically enjoy good health.

Akbash

The Akbash is smaller than most canines used for livestock protection.

Like many other livestock guardian dogs, the Akbash has white fur, albeit varying in length. They're also smaller than most canines used for livestock protection, often slenderer than stout. Still, they're known to be excellent guards against predatory carnivores like wolves. Along with their name, which translates as "white head," these dogs hail from Turkey.

They are independent but require a fair bit of socialization before becoming efficient guardians and good companions to their herd or flock. They'll keep a vigilant watch once they familiarize themselves with their territory and companions. They're non-herding, making them suitable for safeguarding cattle and sheep in a smaller territory. They don't have many health issues and can serve for a long time in most climates.

Tosa Inu

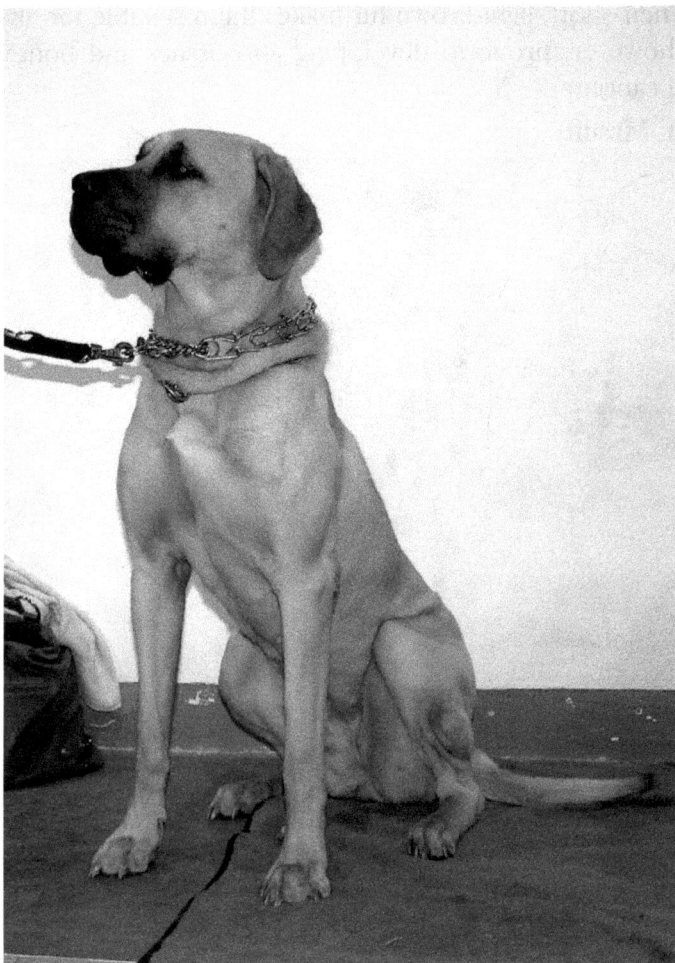

Tosa Inus can reach up to 130 pounds.

Unlike the previous breed, the Tosa Inu is much larger and can reach up to 130 pounds in weight. One of the biggest advantages of a dog this size is that they can become more vigilant as they don't require as much exercise as smaller breeds. Despite their intimidating presence (handy when they need to scare off intruders), they're gentle with their herd or flock. At the same time, their exceptionally high prey drive will ensure that no predator will endanger your livestock or property. Moreover, they require direction, and despite loving to guard passively, they don't like to be without a task.

Unlike other guardian breeds, the Tosa Inu isn't aggressive toward people. Their short, light-brown fur makes them suitable for all climates. They're, however, prone to developing joint issues and bone diseases, along with cancer.

Tibetan Mastiff

The Tibetan Mastiff has an imposing presence.
Alexandr Frolov, CC BY-SA 4.0 <https://creativecommons.org/licenses/by-sa/4.0>, via Wikimedia Commons:
https://commons.wikimedia.org/wiki/File:Tibetan_Mastiff_%D0%A2%D0%B8%D0%B1%D0%B5%D1%82%D1%81%D0%BA%D0%B8%D0%B9_%D0%9C%D0%B0%D1%81%D1%82%D0%B8%D1%84_02.jpg

With their body weight reaching up to a staggering 160 pounds, the Tibetan Mastiff has an imposing presence. Like the previous large canine breed, this one also has a short coat and a short tail. Requiring moderate exercise levels, these make excellent herd dogs for herds and flocks of all sizes. They thrive in colder climates because, despite being short, their fur is relatively dense. However, they can withstand moderately warm temperatures as long as they don't need to run around all day.

Tibetan Mastiffs are incredibly territorial, making them the perfect guard dog. They'll sense intruders from far away and launch into action to protect your farm animals and property if needed. Despite their high prey drive, they're gentle giants when it comes to their herd or flock, and they socialize well with their owners (including small children). They are prone to having eye issues and hereditary conditions like hip and elbow dysplasia.

How to Choose the Right Livestock Guardian Dog Breed

If you are wondering how to choose the right breed based on your needs, livestock, and climate, here are a few tips to help you make an informed decision.

Determine Your Needs

The first step in finding an adequate guard dog is determining your needs based on the animals you keep and where you keep them. Do you have predominantly large livestock, like cattle or horses? If so, you'll need a low-energy dog. However, suppose you have smaller animals or ones that need to be herded. In that case, a high-energy canine will thrive on running around. By contrast, these dogs won't be a good fit for indoor guarding as they'll get bored, inattentive, or aggressive. Also, consider how much space the canine guardians will need to cover. If you have a large farm with multiple pastures, you'll likely need several dogs – in which case, opt for ones that work well in groups. Answering these and other questions regarding your requirements and conditions will help you find the canine that best matches them.

Think about Your Guard Dog's Needs

Another crucial aspect to consider is all your canine guard's needs. While most livestock guardian dogs thrive on being independent workers, some require more attention and care than others. Some dogs can be left alone to guard for days, and you'll need to provide them with enough food and water for this time. These are great if you live away from your farm and can only visit on certain days. Other dogs are more social and thrive better if they interact with their owner daily. Think about your climate. Guard dogs live outside, but if you live in a climate with unstable weather or exceedingly warm temperatures, your canine protector will need some sort of shelter from the elements.

Research the Potential Breeds and Dogs

Once you have your needs and the condition down, you can research the different guard dog breeds to narrow down your choice. Here, consider *training*. Some breeders train guard dogs for exact purposes. However, you'll still need to take care of socializing your canine on your property and with the animals it will be guarding and working alongside (if you have multiple guard dogs). Depending on what type of animals you have, you might also have to think about specialized training for tasks like herding, for example. Consider how much you are willing to invest in training (some dogs require far more training than others) before opting for one breed or another. When you find the right breed, you can proceed to look for the individual dogs. Remember, not all dogs will be suitable for becoming livestock guardians despite their breed being known as protectors.

Is Roaming an Option?

While livestock guardian dogs can be trained to stick to a fenced-in territory, some are more prone to roaming. High-energy dogs will prefer to cover a larger distance when looking for prey, and if you have a small farm, this might take them outside your property. Besides causing potential issues with neighbors, this can also be problematic if they refuse to stay inside the small area you placed them in alongside the herd or flock they are supposed to guard. If you want your canine guardians to remain in small pastures, opt for a non-roaming breed that will be satisfied with passive protection measures, like watching for signs of predators from livestock behavior.

Consider the Canine's Temperament

While most livestock guardians are characterized by a calm temperament, those requiring higher activity levels might become anxious or aggressive in small spaces. At the same time, you don't want your guard dog to be shy or cower in front of other animals, especially if you need them to protect against predators. *Prey drive* is a dog's instinctive characteristic to find, pursue, and capture another animal. Most livestock guardian dogs are bred not to have a strong prey drive so that they can guard farm animals instead of hunting and hurting them. However, some canine guardians will still instinctively pursue other animals that are encroaching or threatening to encroach on their territory. This can be a good thing for keeping predatory carnivores at bay. In this case, investigating whether a threat exists (as most canines with a low prey drive do) isn't enough. They need to actively engage against the intruders. If you live in an area with a high prevalence of predators, you'll need a dog with a high prey drive but one that won't harm your livestock. Here, the goal is to find a balance between calmness and alertness. However, if you don't have many predators in the area and your main concern is to have a herding dog that eventually scares away critters that might damage your property and carry diseases, then you'll need a more mild-mannered dog with a lower prey drive to avoid them biting or chasing the farm animals.

Barking is another factor to consider. Do you want your canine guard to alert you of a threat to your livestock with intense barking but then wait for your orders unless the intruder attacks? If not, then you need an independent thinker who leaps into action instead of just altering with barking or howling. Moreover, barkers might not get along with certain types of animals and might get even more agitated and confused. While being alerted to a threat is beneficial for the guard dog, creating confusion can make them more vulnerable to predators (for example, they might run in the opposite direction than where they are supposed to run and end up in an area where predators can get to them).

Look into the Dog's Interactions with the Livestock

Consider how you want the canine guardians to interact with your livestock. Do you want your stock to follow the dog's guidance when herding but not be afraid of the dog? If yes, look for a breed with a naturally imposing demeanor without being overly dominant. They should be cautious, but it is easy to socialize with farm animals. Some breeds prefer to guard from a distance and not be picked or accidentally stepped on by livestock. If you have to keep larger animals, you don't want a dog that will bite them if they accidentally prod and poke it. Likewise, some dog breeds prefer running around the stock territory, while others walk calmly with your livestock, quietly observing their behavior. The latter will also readily sleep next to the animals, so if this is something you need, look for breeds that don't mind interacting and socializing with their herd flock.

Chapter 5: Training, Feeding, and Care

This chapter will guide you through the fundamentals of LGD training. You'll understand what obedience or communication training is and why it's an essential precursor to socialization. You'll also find LGD training tips, tricks, core commands, and other commands to teach your dog. Then, you'll find out about the different nutrition options you can choose from and understand the factors that determine the type and amount of food your dog needs to thrive. Finally, you'll learn about veterinary and dental care, vaccinations, spotting signs of illness, parasite control, and grooming.

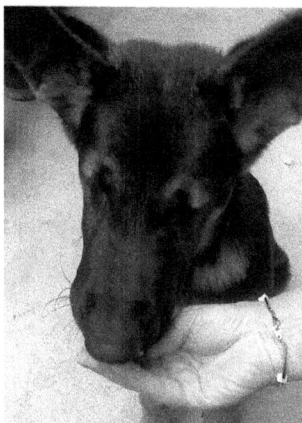

Guardian dogs require proper care.
https://commons.wikimedia.org/wiki/File:German_Shepherd_puppy_eating_out_of_a_human_hand.jpg

Training Your Livestock Guard Dog

The belief that guard dogs don't need to interact and bond with humans is a very common misconception. Like all others, livestock guard dogs crave to build companionship with their humans. While LGDs (livestock guardian dogs) are highly intelligent and capable of protecting livestock, even without constant supervision, they still turn to their owners for support, guidance, approval, and, most importantly, partnership.

Training LGDs is a long process. However, it shouldn't be too hard to do with the right techniques and tools. If you choose the right breed for the intended person, it should respond to your efforts as long as the training sessions are fun, light, and interesting.

Obedience (Communication) Training

When training an LGD, you must start with obedience training before introducing it to the livestock it protects. This will set it up for success in its role as a guard dog and ensure that it responds to your commands and engages in desirable behaviors. Obedience training is integral for healthy connections and interactions and the safety of your dog, livestock, and people.

Guard dogs are usually very large and powerful. If they're not obedience trained, any sudden movement or action they make could put others in danger. Imagine your large Mastiff jumping up on a child (even with the intention of playing with them). If you can't get the dog to listen to you, it would probably hurt that child.

It could also endanger itself if it doesn't listen to your commands. For instance, if the dog attacks a predator and you notice that the animal is trying to get away, the dog won't respond if you command it to "come" or let go. The predator, in that case, could hurt your dog. In some cases, LDGs that have never witnessed birth could kill newborn livestock, mistaking them as potential sources of danger. This is another situation where obedience training comes in handy.

Building a strong bond is both a foundation and an outcome of obedience training. If you want your dog to listen to you, you must invest time and effort into building a strong relationship with it. You must create a partnership where you both care for, respect, and look out for each other. Your dog will reciprocate how you treat it and what you feel for it.

Many dog trainers suggest that "obedience training is a dated term, and that the word 'obedience' should be replaced with 'communication.'" You should never approach your dog from a place of power and control. Instead, think of it as a way to teach your dog how to work with you and an opportunity to learn how to work with it. By maintaining the mindset of partnership and communication, you will eventually get the dog to obey you.

Positive Training Techniques

Positive training methods require you to reinforce desirable behaviors rather than using punishment and shame to deter your dog from undesirable behaviors. Using praise words, play breaks, and treats and making training sessions feel more like play will allow your dog to perceive training and obedience as fun and rewarding rather than dutiful and overbearing.

When training your dog, you should be mindful of its needs and behaviors. For instance, if it's engaging in unusual behaviors or barking more than usual, you need to show concern. Check if there's something it's barking at, pat it, and reassure it. Responding to your dog's needs and concerns will be perceived as a sign of care and respect, further strengthening your bond and enhancing the effectiveness of your training efforts. Not only is using positive training methods highly effective, but it's also the only ethical and humane way to treat your dog.

Tips and Tricks for Training Your Dog

1. **Release All Pent-Up Energy**. If your dog is excited and all worked up during the training session, then you're going to struggle to teach them anything. Ensure they've had enough time to exercise and play so they're calmer and less hyper when you start.

2. **Create a Distraction-Free Zone**. Most dogs are very easily distracted by any sound, object, or even smell that they find interesting. Dogs, especially those that have never been trained, usually find it hard to learn new things. Training them in a place where they're likely to encounter various external stimuli doesn't make it any easier for them. Choose a distraction-free environment for your training sessions for more fruitful learning experiences.

3. **Most Importantly, Enjoy.** Dogs are naturally playful creatures. They respond and engage with things that they perceive as fun. Your training sessions should be fun and feel like play rather than a chore. Use this as an opportunity to bond with your dog and spend some quality, fun time together.

4. **Set Up a Reward System.** You need to use rewards to reinforce good behavior by giving the dog something it wants whenever it responds to one of your cues. Getting to know your dog allows you to determine which reward it wants the most. Some dogs prefer treats, some love to play with their favorite toy for a few minutes, and for others, pats, belly rubs, and a few praise words will do the trick.

5. **Know When Breaks Are Needed.** Chances are that your efforts will backfire if you resume training your dog when you're both clearly frustrated. Sometimes, both of you should take a break for a few days. When you're on break, however, avoid making any cues or commands to your dog; if it doesn't respond, you'll have to return to training it. You should also always try to end your session with success.

6. **Try Your Commands in Different Settings.** If you've successfully gotten your dog to "come," for instance, in the backyard, you have to try to get it to respond to the command in the park, on the porch, or inside the house. Some dogs don't understand that their response to a given command doesn't depend on the context or setting, so you should test out their responses in different places.

7. **Introduce Distractions.** If your dog successfully responds to your commands in different settings, you need to gradually introduce distractions to the context. You must ensure that your dog listens to you even when it finds something more interesting to do or explore. Remember that as a livestock guard, your dog will be surrounded by multiple distractions, so this step is essential. Once you complete its basic obedience training, you have to work on training it in a busy park, around other people and animals, and around livestock. You should also train it even when it's still hyperactive and hasn't had the chance to release its excess energy. Your training is successful when your dog responds to you no matter how it feels and regardless of its

surroundings.

The duration of the training depends on how responsive your dog is. However, in most cases, you'll have to consistently train your dog for several months before it's ready to guard livestock. Don't give up, and remember that no matter how impossible it may feel, all dogs are trainable if you invest the time, effort, and consistency they need.

Expect your dog's learning pace to vary across different commands. It may pick up on some in just a few days and take weeks to learn from others. Avoiding frustration and staying patient is the key. You should also be mindful of when you stop or decrease the rewards. You have to do it when you're certain you won't lose your ground.

The 6 Core Commands

Several commands come in handy for a livestock guard dog; however, the 6 most important ones are:

1. **Look at Me.** This command is very helpful and usually easy for dogs to learn. It can set you up for success in the rest of your training because it encourages your dog to look at you and focus on what you have to say. You can use this command as a distraction if you need it to stop doing something it shouldn't do. It also allows you to (usually) get its attention regardless of the situation.

2. **Sit.** This command is necessary for your dog to learn before you can teach it to "stay." It is easier for dogs to learn to stay when they're already sitting down.

3. **Stay.** Teaching your dog to stay is crucial because, in many cases, it's the only way you can get it to back away for its safety or stop it from hurting others.

4. **Come.** Like the "stay" command, "come" can be used to stop your dog from harming itself or others. It is often a safer solution to the problem as well.

5. **Leave It.** This is another command that can both protect your dog and the livestock. If you notice a predator trying to get away or your dog mistakenly attacks livestock (newborn or new livestock), you can safely ask it to leave the animal alone.

6. **Off.** Another easy command you can teach your dog is asking them to get off if they jump on you or others. Many dogs stop

jumping on people entirely if their owners repeat this command enough for them to understand that this is something that they shouldn't be doing.

Other helpful but not essential commands include stand, down, and shake. Some people also ask their dogs to bark on demand through a "speak" command. This command can actually help enhance the dog's protective instincts, encouraging it to bark when something goes wrong.

Duration and Challenges

While there are general guidelines you need to follow when training your LGD, you have to keep in mind that the training process is not one-size-fits-all. Like humans, each dog has a unique personality that makes it more responsive to certain types of training on command. While some friendlier and naturally protective dogs are more inclined to bond with and guard livestock, others might need more time to understand what's asked of them. This also applies to your livestock. Some will easily adapt to and respond well to the dog's presence. In contrast, others will initially show signs of aggression toward the dog.

Every livestock and LGD owner experiences unique challenges throughout the process because the situation solely depends on how the animals deal with each other. The socialization process is the most crucial step. It is where most problems arise and when you have to be the most patient and creative in finding solutions. You should also know that the training process requires a lot of trial and error until you figure out what works best for your dog.

The best part is that if you decide to acquire and train more dogs later on, you will face significantly fewer challenges. You will be more experienced and prepared, and the puppies will be more responsive because they'll naturally model the trained LGD's behaviors. In many cases, the older LGDs correct the younger puppies, making your only responsibility overseeing and monitoring the process.

When thinking of getting an LGD, you should also consider getting more than one dog. LGDs work best in packs, are likely to exhibit significantly fewer behavioral issues and experience training challenges when working in groups. Most importantly, one dog cannot effectively control predators and protect your entire ranch. The exact number of guard animals you need, however, depends on the size of your ranch, the number and type of livestock you own, and the nature of surrounding predators.

Feeding Your LGD

Suppose you've been reading into LGD care and feeding for a while. In that case, you've probably noticed that the greatest debate regarding LGD nutrition is whether you should opt for commercial or raw food. Many professionals suggest that LGDs should be fed raw food since they're descendants of wolves. In contrast, others argue that domesticated dogs have entirely different dietary needs and digestive systems, calling for a more varied and balanced diet. There is no solid right or wrong approach to how you should feed your dog as long as it's healthy and getting all the nutrients it needs to thrive.

Commercial Dog Food

Most LGD owners prefer to feed their dogs high-quality commercial food (some people also recommend grain-free types), as it contains the right balance of proteins, fats, carbohydrates, vitamins, and minerals. This is a good place to start, and you can always change your mind and try different diets further down the road. Make sure, however, to consult a vet before making any substantial differences.

Numerous commercial dog food brands are on the market, ranging from highly affordable to extremely expensive choices. Unfortunately, dog owners have found time and time again that you get what you pay for in terms of quality and nutritional value. Cheap brands will not satisfy your dog's nutritional needs, always leaving it hungry and fatigued. You'll end up having to feed it more, which can cause the dog to become overweight. Even when this happens, your dog will still be nutritionally deprived. Use very large and durable bowls to feed your dog.

Feeding Amounts

You can switch between a few high-quality brands so your dog doesn't get bored after a while. Adult dogs should generally be fed twice daily, while puppies require an additional meal. That said, some dog owners prefer to feed puppies around 4 times a day until they're at least three months old. While these are generic insights, you should consult your vet because they'll better understand your dog's unique needs. The exact amount of food that a dog needs depends on their weight and activity levels. Older, pregnant, and lactating dogs and those with health conditions require specific feeding requirements, too.

If you have many dogs on your ranch, feeding each of them twice a day can be a very daunting task. This is why many ranch owners prefer to free-feed their LGDs. If you must do that as well, try to keep a watchful eye on your dogs' eating habits, as well as their weight, to make sure that they're not putting on more weight than what is considered healthy.

If you plan on putting your dog on a raw diet, you should also consult your vet on how to safely implement it and ensure that your dog meets its nutritional needs. Some brands offer frozen, dried, and dehydrated raw food options for dogs, which are considered convenient and easy to serve.

Healthcare and Routine Care

Veterinary Care

- Adult dogs should get an annual veterinary examination (at least).
- Puppies (up to 4 months old) should see the vet every 3 to 4 weeks.
- Dogs over 7 years old should get at least 2 veterinary examinations a year.

Signs of Illness

Dog owners can easily sport signs of illness by observing changes in their dog's behavior. You should take your dog to the vet if you notice changes in appetite, activity levels, urination frequency, itchiness, lameness, or more obvious signs of illness.

Vaccinations

Dogs have to be vaccinated just as people do. Follow up with your vet to stay on top of your dog's routine vaccinations to protect it against diseases. Rabies, distemper, and parvovirus are considered core vaccines. However, depending on the environmental and regional circumstances, your dog might need additional vaccines.

Parasite Control

Tapeworms, roundworms, whipworms, and hookworms are among the main intestinal parasites that dogs can develop. These parasites can severely damage the animal's digestive system and hinder its ability to absorb the nutrients it needs. The best way to detect whether your dog has an intestinal parasite is to get a fecal sample tested at least once a year. Make sure to clean up your dog's feces frequently, as some parasites can be transferred to humans and other animals.

Some parasites, like heartworms, which are often fatal, are transmitted through mosquito bites. Dogs can also contract external parasites like mange mites, fleas, and ticks. You can use specialized body sprays as preventative measures to protect your dog.

Dental Care

Feeding your dog dry food, offering certain chew toys, and brushing its teeth regularly will help keep its gums and teeth healthy.

Grooming

You should brush your dog's hair regularly to get rid of tangles, remove small objects that might've accidentally gotten caught up in it, and remove shed hair. Mats and tangles can irritate your dog's skin, making them more likely to develop infections. You should also bathe your dog with pet shampoo periodically. Consult your vet regarding how often you need to wash your dog, as excessive washing can irritate skin.

This chapter serves as a mini-guide on how to train, feed, and care for livestock guardian dogs to ensure their well-being and effectiveness in protecting your livestock. Now that you've read it, you understand why training, nutrition, and care are crucial for your dog's success in their guardian role.

Chapter 6: Troubleshooting LGD Behaviors

In this pivotal chapter, you'll explore Livestock Guardian Dog (LGD) behaviors, recognizing them as the linchpin in adequately protecting livestock and property. The emphasis here extends beyond addressing behavioral challenges solely for the functionality of LGDs; it resonates with the fundamental understanding that their well-being is inexorably linked to their proficiency as guardians. This chapter navigates the identification and understanding of behavioral issues. It intricately outlines methods to effectively address and resolve these challenges.

Take the time to understand your guardian dog's behavior.
https://unsplash.com/photos/woman-hugging-a-dog-FtyJIuBbUhI

Conducting a Thorough Assessment

When navigating Livestock Guardian Dog behaviors, conducting an assessment is crucial. This multifaceted process involves keen observation, baseline establishment, trigger identification, health evaluation, and analysis of interactions with both humans and livestock. Here are the steps you can take to conduct a thorough assessment.

Observational Techniques

To initiate this process, observe the LGD in its natural environment. Witness the intricate dance of its interactions with livestock, noting its demeanor and actions. Observe how it responds to potential threats, displaying alertness, aggression, or fear. Document regular behavioral patterns, including territorial marking, patrolling routes, and social interactions with humans and other animals.

Establishing Baseline Behaviors

Creating a baseline is pivotal for identifying deviations that may signal underlying challenges. Document the LGD's typical daily activities, including patrolling, resting, and interacting with livestock and humans. Understand the dog's sense of territory, noting its territorial behaviors and how it marks and defends its space. This baseline serves as a reference point for recognizing changes in behavior.

Identifying Triggers

Triggers play a significant role in influencing LGD behavior. Identify environmental changes that might act as triggers, such as new wildlife, construction, or alterations in the landscape. Record the LGD's reactions to these potential triggers, noting whether it responds calmly or displays signs of stress or aggression. This understanding aids in addressing the root causes of undesirable behaviors.

Health and Physical Examination

The integration of health into the assessment process is crucial. Collaborate with veterinary professionals to conduct a thorough physical examination, ensuring the LGD is in optimal physical health. Address any health issues that might influence its behavior, recognizing that pain or discomfort can manifest behavioral changes. A holistic approach encompasses both physical and mental well-being.

Analyzing Interactions

Evaluate how the LGD interacts with both humans and livestock. Assess the bond quality between the LGD and its caregivers, ensuring consistent and positive communication. Examine the LGD's interactions with livestock, looking for signs of stress or aggression and evaluating its effectiveness in protecting the herd. The dynamics of these interactions provide profound insights into the LGD's performance and challenges.

Behavioral Deviations

Identifying deviations from established baselines is vital. Record any changes in behavior, such as increased aggression, withdrawal, or altered patrolling patterns. Document when these behavioral deviations occur – whether related to specific times of the day, seasons, or events. This nuanced understanding helps pinpoint challenges and aids in tailoring targeted intervention strategies.

Developing a Customized Intervention Plan

Upon identifying challenges, empower yourself to develop a tailored intervention plan. Address each identified challenge individually, focusing on triggers and underlying causes. Emphasize gradual interventions to avoid overwhelming the LGD, ensuring consistency in reinforcing positive behaviors. This customized approach acknowledges the uniqueness of each dog and situation.

Holistic Well-Being

Recognize the importance of holistic well-being for LGDs. Consider mental stimulation as a vital component, including activities that engage their intelligence and instincts. Create an environment that fosters emotional fulfillment, ensuring a content and balanced LGD. Holistic well-being contributes to adequate protection and the overall quality of life for the LGD.

Ongoing Monitoring

Highlight the dynamic nature of LGD behaviors and the need for ongoing monitoring. Encourage continuous assessment to track the effectiveness of interventions and note any evolving behavioral patterns. Stress the importance of adaptability in strategies, as LGD behaviors may evolve, requiring adjustments to the intervention plan. This cyclical approach ensures sustained effectiveness and a responsive guardian.

Common Behavioral Issues

• Aggression Toward Humans

Territorial Instincts

Livestock Guardian Dogs (LGDs) exhibit a robust territorial instinct rooted in their evolutionary history as protectors. This behavior manifests as aggression towards humans perceived as intruders in their designated territory. This instinct is vital to their protective nature but requires careful management to ensure appropriate boundaries.

Fear or Anxiety

Aggression stemming from fear or anxiety is a complex aspect of LGD behavior. It often traces back to insufficient socialization during the critical developmental stages. LGDs not exposed to various environments, people, and stimuli may perceive humans as threats, triggering defensive aggression. Understanding these fear-based responses is crucial for addressing the root cause.

Protecting Resources

Resource guarding, including aggression to protect valuable resources, is ingrained in LGDs. This can extend to food, shelter, and favored areas within their territory. Recognizing and managing resource-based aggression involves creating an environment that minimizes competition and stress and develops a harmonious coexistence between the LGD and humans.

• Aggression Toward Livestock

Poor Socialization

Inadequate socialization during the formative stages of an LGD's life can result in aggression towards livestock. Without exposure to various animals during their early development, LGDs may view livestock as unfamiliar and potentially threatening. Addressing poor socialization involves gradual introductions and positive reinforcement to build trust.

Stress or Fear

Changes in the environment, handling procedures, or unexpected events can induce stress or fear in LGDs, leading to aggression towards livestock. Understanding the triggers and sources of stress is essential for developing strategies that alleviate anxiety, creating an environment where the LGD feels secure in its role as a guardian.

Maternal Instincts

Female LGDs may display protective aggression, particularly during breeding seasons or when safeguarding their young. This maternal instinct is a natural behavior rooted in ensuring the safety and well-being of the herd. Managing aggression related to maternal instincts involves providing a supportive environment in order for the LGD to fulfill its protective role without compromising safety.

• Barking Behavior
Territorial Defense

Excessive barking is a tool for territorial defense – a fundamental aspect of an LGD's role. Understanding that barking is a form of communication to establish and defend territory is crucial. Effective interventions allow the LGD to fulfill its protective duties and minimize excessive noise in residential or shared environments.

Lack of Stimulation

Boredom is a common trigger for excessive barking in LGDs. These intelligent and active dogs require mental and physical stimulation. Implementing enrichment activities, interactive play, and varied stimuli can alleviate boredom, address the root cause of the excessive barking, and promote a more contented LGD.

Communication

Barking is a natural means of communication for LGDs. Increased barking may signify heightened awareness of perceived threats or a need to communicate with humans or other animals. Deciphering the nuances of their communication cues enables owners to respond appropriately, fostering a more profound understanding between humans and LGDs.

• Predator Apathy
Lack of Training

Apathy towards potential predators can result from insufficient training. LGDs need to be trained to recognize and respond to specific threats. Implementing consistent and positive reinforcement training techniques enhances their ability to discern between friend and foe, ensuring effective predator control.

Health Issues

Physical discomfort or health problems may compromise an LGD's ability to perform its guardian duties effectively. Apathy may be a symptom of an underlying health issue. Regular veterinary checkups, promptly addressing health concerns, and adjusting the LGD's responsibilities during illness contribute to overall well-being.

Overfamiliarity with Livestock

LGDs raised alongside livestock from an early age may become less responsive to potential threats, especially if they perceive the livestock as part of their social group. Overfamiliarity involves balancing companionship with a vigilant stance towards potential predators, preserving the LGD's protective instincts.

• Roaming Behavior

Lack of Boundaries

Roaming behavior often results from inadequate fencing or unclear territorial boundaries. Establishing clear boundaries through secure fencing is crucial. Creating a defined territory helps the LGD understand its limits, reducing the likelihood of wandering and promoting effective territorial control.

Hormonal Influences

Unneutered or unspayed LGDs may roam in search of mates during breeding seasons. Addressing hormonal influences through spaying/neutering is a proactive measure to mitigate wandering behavior, ensuring that the LGD remains focused on its protective responsibilities.

Socialization Issues

Dogs that lack proper socialization may roam in search of companionship or stimuli. Introducing positive socialization experiences and ensuring that the LGD has adequate interaction with humans and other animals help address socialization issues, reducing the motivation to roam for social fulfillment.

• Fearful Behavior

Lack of Socialization

Fearful or timid behavior often stems from insufficient exposure to diverse environments, people, and stimuli during the critical socialization period in a dog's early life. Gradual exposure to positive experiences and a supportive environment aids in building confidence and addressing the root cause of fear-based behaviors.

Traumatic Experiences

Past negative encounters, abuse, or traumatic experiences can lead to persistent fear and timid behavior. Rehabilitation involves rebuilding trust through patient and positive interactions. Creating a safe and predictable environment supports the emotional well-being of LGDs recovering from traumatic experiences.

Genetic Predisposition

Some LGDs may have a genetic predisposition toward timidity. While genetic factors play a role, proactive socialization and positive reinforcement can help mitigate inherent tendencies. Understanding the interplay between genetics and environmental factors is essential for tailoring behavioral interventions.

Anxiety or Stress

Changes in the environment, routine, or the introduction of new elements can induce anxiety or stress, leading to increased marking behavior. Addressing the root cause involves identifying stress triggers and implementing measures to create a stable and secure environment, promoting emotional well-being.

Hormonal Influences

Intact males may engage in more frequent marking behavior, especially during breeding seasons, influenced by hormonal changes in reproductive activities. Spaying/neutering can mitigate hormonal influences, providing a more consistent and controllable behavior pattern.

Understanding these common behavioral issues and their nuanced root causes is pivotal for implementing effective intervention strategies. Tailoring solutions to address the specific underlying issues contributes to behavior modification and the overall well-being and effectiveness of Livestock Guardian Dogs in their protective roles.

Behavior Modification Techniques

Initiating behavior modification begins with a thorough assessment. Observe the specific behaviors causing concern, identifying triggers, contexts, and patterns. Documenting these observations provides a baseline for addressing challenges.

Positive Reinforcement Training

Utilize positive reinforcement to encourage desired behaviors. Reward the LGD for exhibiting appropriate responses to stimuli or situations. Consistency in rewarding positive behavior fosters a deeper understanding of expectations.

Desensitization

Gradually expose the LGD to stimuli that trigger undesired behaviors - pair exposure with positive experiences or rewards to create positive associations. Slowly increase intensity or proximity as the LGD becomes more comfortable, reducing adverse reactions.

Establish Clear Communication

Consistent verbal and visual cues are essential for effective communication. Reinforce commands consistently, ensuring that the LGD understands and responds reliably. Clear communication enhances the bond between the owner and the LGD.

Controlled Socialization

Gradual exposure to new people, animals, and environments helps build confidence and reduce fear. Facilitate positive interactions during socialization to create positive associations. Monitor reactions and adjust the socialization pace based on the LGD's comfort level.

Environmental Enrichment

Provide mental and physical stimulation through toys, puzzles, and interactive activities. Rotate toys regularly to maintain novelty and prevent boredom. Engage in activities that simulate the LGD's instincts, such as scent work or puzzle-solving.

Consistent Boundaries

Clearly define and reinforce territorial boundaries. Use fencing and visual cues to help the LGD understand its limits. Reward adherence to boundaries with positive reinforcement, reinforcing a sense of security.

Addressing Fear-Based Behaviors

Identify and avoid triggers that induce fear. Gradual exposure to fear-inducing stimuli and reassurance and rewards for calm behavior reduce fear-based behaviors. A supportive environment is crucial in addressing fear.

Professional Guidance

Seeking assistance from experienced trainers or behaviorists for specific challenges is valuable. Professionals can provide tailored strategies based on the LGD's needs, ensuring a comprehensive and informed approach to behavior modification.

Behavioral Analysis

Conduct a detailed analysis of the LGD's behavior, considering past experiences, traumas, or environmental changes. Understanding the root cause facilitates targeted interventions.

Use of Deterrents

Introduce deterrents for undesirable behaviors, such as excessive barking. Devices emitting sounds or vibrations can discourage behaviors without causing harm.

Interactive Training Games

Incorporate interactive training games to stimulate the LGD mentally. Puzzle toys or games that require problem-solving engage their cognitive abilities.

Variety in Commands

Introduce a variety of commands to keep training sessions interesting. This prevents monotony and ensures the LGD remains attentive and responsive.

Scheduled Playtime

Designate specific periods for playtime and interaction. Structured play helps release excess energy and reinforces positive behavior.

Calming Techniques

Learn and implement calming techniques, such as a massage or gentle stroking, to soothe the LGD during stressful situations.

Preventive Measures

Begin socialization during the critical developmental period. Introduce the LGD to various environments, people, and animals to prevent fear-based behaviors from developing.

Environmental Assessments

Regularly assess the LGD's living environment for potential stressors or changes. Minimizing environmental stressors contributes to stable behavior.

Supervised Socialization Events

Arrange supervised socialization events with other dogs or animals. Controlled interactions provide positive experiences and enhance social skills.

Consistent Exposure

Regularly expose the LGD to various sounds, sights, and smells to prevent fear or anxiety associated with novelty.

Positive Handling Techniques

Implement positive handling techniques to build trust. Gradual acclimatization to handling ensures the LGD remains comfortable in various situations.

Routine Mental Challenges

Create routine mental challenges, such as hiding treats or toys, to stimulate the LGD's problem-solving abilities and mental agility.

Adjustable Training Intensity

Modify training intensity based on the LGD's responsiveness and energy levels. Tailoring the training regimen prevents fatigue and maintains enthusiasm.

Diversified Play Equipment

Introduce a variety of play equipment, like agility obstacles, to keep physical activities diverse and enjoyable for the LGD.

Consistent Verbal Reinforcement

Use consistent verbal reinforcement throughout the day to acknowledge positive behavior. Verbal cues contribute to reinforcing established commands.

Sensory Stimulation

Incorporate sensory stimulation through different textures, surfaces, and environments. Engaging multiple senses enriches the LGD's overall experience.

Structured Rest Periods

Establish structured rest periods within the daily routine. Adequate rest supports mental well-being and behavioral stability.

Flexible Training Schedule

Maintain flexibility in the training schedule to accommodate the LGD's energy levels and adaptability. A flexible approach ensures training remains enjoyable.

Regular Exercise

Provide ample opportunities for physical exercise in order to prevent boredom and excess energy. A well-exercised LGD is less likely to engage in undesirable behaviors.

Health Checkups

Schedule regular veterinary checkups to address any potential health issues. Physical discomfort can contribute to behavioral problems, so maintaining good health is essential.

Consultation with Professionals

Consult with experienced trainers or behaviorists for guidance. Professionals can offer personalized insights and strategies based on the LGD's unique characteristics, ensuring a holistic approach to behavior management.

Implementing these behavior modification techniques and preventive measures creates a comprehensive strategy for fostering positive behavior in Livestock Guardian Dogs. Recognizing the individuality of each LGD and making adjustments based on their responses and progress is essential for successful behavior modification. Seeking professional guidance adds an extra layer of expertise to addressing specific challenges effectively.

Ethical Responsibility

Livestock Guardian Dogs (LGDs) are indispensable protectors, but their well-being and humane treatment are equally paramount. Responsible LGD ownership entails a commitment to lifelong care. Recognizing the ethical responsibilities of LGD ownership ensures a holistic and humane approach.

Commitment to Lifelong Well-being

Owning an LGD is a lifelong commitment. As guardians, owners must pledge continuous care, addressing their companions' evolving physical and emotional needs throughout their lives.

Ethical Breeding Practices

Supporting ethical breeding practices is essential. This involves prioritizing health and temperament over profit and avoiding patronizing breeders who compromise the well-being of their dogs.

Adequate Living Conditions

Providing suitable living conditions is fundamental. LGDs deserve shelter, comfortable bedding, and a designated space that caters to their needs, promoting a sense of security.

Balanced Nutrition

Prioritize a balanced and nutritious diet. Collaborate with veterinarians to determine the optimal diet based on age, size, and activity level, ensuring their overall health.

Regular Veterinary Care

Scheduled veterinary checkups are non-negotiable. Regular visits allow for preventive care and early intervention, safeguarding the LGD's health and well-being.

Mental Stimulation

Recognize their intelligence and working nature. Provide mental stimulation through activities, toys, and interactions, preventing boredom and promoting mental well-being.

Socialization and Positive Reinforcement

Prioritize positive reinforcement and gentle training techniques. Continuous socialization efforts contribute to a well-adjusted and confident LGD, enhancing their overall welfare.

Prevention of Cruelty and Neglect

Uphold a zero-tolerance policy for cruelty and neglect. LGD owners must actively report signs of abuse or neglect within the community to protect all dogs' welfare.

Respect for Natural Behaviors

Acknowledge and respect the dog's natural behaviors. Avoid punitive measures that may compromise their well-being, fostering a harmonious, ethical owner-dog relationship.

Adaptation to Individual Needs

Recognize that each LGD is an individual with unique needs. Tailor care and training strategies to accommodate their distinct personalities, ensuring a customized approach to their well-being.

Consideration of Retirement

Plan for the LGD's retirement. Ensure a comfortable and peaceful transition as they age, acknowledging their contributions and providing for their changing needs.

Integration into Family Life

Integrate LGDs beyond their working duties into family life. Fostering a bond based on companionship and mutual respect enhances their overall well-being.

Humane End-of-Life Decisions

Face end-of-life decisions with compassion. Make decisions prioritizing the LGD's comfort, minimizing suffering during their final stages, and ensuring a dignified farewell.

Ethical LGD ownership encompasses a profound commitment to these remarkable animals' welfare, health, and happiness. Responsible owners are pivotal in setting ethical standards within the broader LGD community.

Chapter 7: Llamas as Guardians

If someone told you that a llama watches their sheep, you might think it is a dry, absurd dad joke. There is no need to force a chuckle because they are probably serious, but you can wait for the punchline if you'd like. Don't be fooled by the llama's cute and awkward appeal because these animals make brilliant guardians. They are not what you'd typically believe a guardian animal to be because they are grazers, not predators. However, llamas are powerful and territorial creatures that form deep bonds with the livestock that they protect. Furthermore, since llamas aren't predators, you do not have to worry about them mauling your livestock like cats or dogs.

Llamas can graze alongside many species, increasing their time under their watchful eyes. In the right conditions, llamas are the perfect caretakers. Llamas are not excessively expensive and are relatively easy to maintain. Considering their effectiveness as guardians against numerous predator species, your pockets will be smiling at you. Their bulky, fur-covered, 400-pound body towering over your other animals is like a walking battle station looking to bite, kick, and stomp any intruder that crosses the line. This fox chaser will have the unwanted snackers running for the hills at top speed to avoid their ferociousness.

Llamas can be guardians.
*Ralf Roletschek, GFDL 1.2 <http://www.gnu.org/licenses/old-licenses/fdl-1.2.html>, via
Wikimedia Commons. https://commons.wikimedia.org/wiki/File:18-08-25-
%C3%85land_RRK6596a.jpg*

Llamas are generally calm and trainable, so you can keep them around children if you are careful, alert, and informed. They can develop a bad temper if they are mistreated, so you must care for them ethically for your safety and their well-being. Llamas, especially when young, are susceptible to many diseases, so you must ensure that you take the right medical precautions while remaining vigilant of signs of infection. Your fuzzy companion can keep you entertained for hours with their good looks and affection as a pack animal while they stay on duty, keeping your livestock guarded. So, give these cuties with their trendy haircuts a chance because their competence as a guardian may surprise you.

Physical and Behavioral Characteristics

Seeing a llama in a picture and seeing one in real life are two different experiences. The body of a llama is about four feet tall, but their long neck takes them up to about six feet. They can weigh anything from about 250 to 450 pounds, but generally, like with many other mammal species, the females are smaller. This is why people opt for sterilized males because they are bigger and less aggressive than a testosterone-filled giant.

Llamas are brilliant livestock guardians because they are naturally aggressive towards foxes, coyotes, and dogs, which are common enemies for poultry, sheep, and goats. Having their eyes on the side of their head gives them a 360 view that can spot a predator approaching from a distance. They thrive in high-altitude environments because of the hemoglobin levels of their blood and can run up to 40 mph, so you will not beat them in a foot race. The diseases they develop are usually nutrition-based because they have strong immune systems to fight off many contagious pathogens. Therefore, what you feed them is crucial to get the most out of your llamas.

Llamas are known for spitting, but they rarely aim at people. This behavior is usually done as a dominance-establishing competition between males trying to be the alpha or when they are irritated. Llamas do not attempt to escape their enclosures; they eat the same food as sheep or goats. They also prefer structured shelters, like those of sheep, which is why they are so well-suited for each other. Llamas are used in the mountains of South America because they can carry a lot of weight up steep hills. Their padded feet have two toes that do not cause as much damage as the hooves of horses or donkeys do.

They live for about 15 to 20 years, so you can trust that they will be around and active for a long time. You do not have to worry about constantly replacing them or even breeding them. Despite their goofy appearance, llamas are athletic and can easily chase down a predator when they need to. They have long, pointy ears used to listen carefully to slight changes in their surroundings as well as to communicate their emotions or temperament.

Llamas have tough fur that protects against injuries and attacks from smaller predators. Although they look similar to alpacas, they are not quite as gentle and are also bigger. Farmers in South America often keep a few llamas in their alpaca herd to fight off unwelcomed guests. They anger quicker than alpacas and do not run away from a confrontation. Unlike dogs, llamas naturally take on the guardian role. Dogs need to be trained so they don't attack the herd, but llamas just fall in line without any instruction. When lambs are born, you'll notice that your llama will take on the babysitter role without being prompted. They are herd animals, so these collective behaviors are expected.

Not many animals are as uniquely suited for the guardian role as llamas. Their physical structure, coupled with their mentality, makes them overachievers as guards. They have six sharp and angled teeth that can do horrific damage if they penetrate flesh. Their long legs keep predators away from their vitals, and they can quickly trample them before they attempt to attack. They are brave, and in some cases, they will even charge at bigger animals, scaring them off with sheer audacity.

Role as Livestock Guardians

Llamas have two main weapons of defense: their huge bodies and shrill screams. As strong, fast, and agile as llamas are, they simply cannot fight everything. However, even in the cases where they see a clear loss coming, they will loudly neigh, alerting anybody within earshot of the pressing danger. Their fancy footwork makes easy work of foxes and coyotes, while their protective nature ensures no one is left behind.

The effectiveness of llamas as guardians can't be denied. A study was conducted by Iowa State University on 145 sheep farms in the United States (Franklin et al., 2012). The study measured the effectiveness of introducing llamas as guardian animals. There was a significant difference between when the llamas arrived and before they got there. The number of attacks on sheep dropped from 11% to 1%, with 50% of farmers claiming that they had experienced no attacks at all (Franklin et al., 2012). The llamas used a variety of strategies, like herding the sheep away from danger or charging at the predators, which included coyotes, dogs, foxes, and even bears.

The role of a guardian llama is to spot danger, warn the herd about an approaching threat, get the herd away from the danger, and, if need be, fight the predator. They have similar diets to sheep, so they will feed alongside your animals. They can be taken out into the pasture to graze with your flock. Unlike guard dogs that do not become part of the flock, as a grazer itself, the llama will insert itself into the group as somewhat of a fatherly figure, protecting its weaker children.

With its all-around vision, powerful sense of smell, and amazing hearing, your llama is there to keep a lookout for predators. They are most effective against medium-sized threats like dogs, foxes, and coyotes but do not do well against bigger predators like bears and lions. Their height of six feet is also an advantage because they can see further than the sheep and goats they typically protect.

Since they will spend a lot of time feeding and in the same area as your livestock, they will establish a natural connection with them that is difficult to break. You do not have to train them like you would a dog, but you can take some steps to maximize their effectiveness. For example, to ensure they establish a bond with your livestock, which makes them protective, you should place them in an adjacent enclosure first and then gradually introduce them to your flock.

The best animals to pair llamas with are sheep or goats. They do not match well with cows because the predators they fight against are not much of a threat to older cows and would usually target calves. A cow can provide the same protection to its young as a llama. They can still assist with chasing predators off and acting as a vigilant alarm. Still, they are just more suited for smaller livestock. Sheep are low-strata grazers, going deep into the soil to eat what the livestock may leave behind, while llamas are medium-strata grazers that take the top half of the grass, as well as browsers that eat from trees and shrubs, so they are not in direct competition for food with sheep.

Taking care of your llama will help it become the best protector it can be. For instance, llamas are sensitive to excessive heat, so you must ensure that there are many shady areas for them to rest in and that they have water to cool themselves down. You need to trim their toes so that they do not get foot rot, and they require regular sheering, which also helps manage the heat. Llamas are susceptible to many parasites, so you have to take the necessary precautions to prevent infection. They also need annual vaccinations against Clostridial Disease, delivered as a five-in-one dose. The meningeal worm is deadly for llamas, so you must deworm them regularly. This parasite is found in deer, so if you have a large deer population in your area, llamas may be the wrong choice for you unless you take precautions to keep deer away.

Pros and Cons of Using Llamas as Guardians

Environmental analysis is the first stage of determining whether llamas are the guardian animals for you. You must think about which kinds of animals you have and how the llama would function within that environment. A lofty, mountainous region is a tick in the llama's favor because they survive well in high altitudes. Llamas become attached to the animals they protect because they are driven by a natural urge. However, what they see as dangerous could affect how appropriate they are for the services you need. For example, llamas will ward off foxes and coyotes but will ignore the much smaller raccoons, skunks, and weasels, which is terrible if you are farming with chickens. Therefore, one of the major downsides of having a llama as a guard is that they do not gel with all kinds of livestock. Dogs may eat chickens, but you can train them to leave them alone and project them, whereas a llama will just ignore the threats to chickens without realizing they are putting the flock in danger.

A huge pro of keeping llamas on your farm is that they adapt to a multitude of climates. The hardy animals do not get hurt easily and will not succumb to weather changes. This resilience to climate, paired with how long they live, means the gorgeous beast will be in service for many years. Younger llamas are not suited as guard animals, so you should get one that is at least over two years old. Preferably, it should be a male that is neutered. Llamas function best as guardians when they are solo, but depending on the size of your flock, you may need more than one.

Llamas are an ethical option for predator control. You may ask how this makes sense since llamas will kill foxes, dogs, and coyotes without hesitation. However, the presence of a llama is usually enough to keep wildlife away from your animals, which means you will be killing a lot fewer foxes than if you had used less ethical measures like poison. Furthermore, poison can inadvertently harm creatures that you do not want to harm, and some foxes have become so smart that they avoid poison bait. Additionally, poison can deteriorate the soil, which, over time, will impact the indigenous plant life on your farm, throwing off the delicate balance of the natural ecosystem.

The similarity of the llama's diet with a sheep or a goat diet means that you can seamlessly introduce them into your flock without much maintenance when it comes to nutritional considerations. However, you must ensure that they have enough copper, zinc, and vitamin D in their diets if you do not want them to develop malnutrition-related illnesses. Llamas don't easily pick up contagious infections, so feeding them properly is half the battle. Furthermore, they should always have access to an ample supply of water to prevent urinary blockages or infections.

Legal Restrictions and Regulations

Llamas, much like dogs, are one of the oldest domesticated animals. In South America, some people say that there is evidence that these gorgeous animals have been living side by side with humans for over 4,000 years. This is why llamas are considered domesticated animals, like sheep and cows, for the most part. However, it is important to be mindful of what is legal in your region. Different laws and ethical standards are applied according to the country or state that you live in. Familiarizing yourself with these regulatory standards will ensure that you keep the authorities at bay and do not have to pay fines, lose your animals, or even spend time in prison.

Sometimes, the laws and regulations put on you when holding animals can seem like a lot. However, it is essential to remember that the bodies that create and enforce rules are there to protect you and your animals. Having standardized regulations ensures that animals will be provided with appropriate care while protecting humans from a variety of dangers, like zoonotic diseases. People's ethical standards differ depending on their morals, values, upbringing, and culture. Therefore, people draw the line for what is appropriate, safe, and cruel at different points. Individuals can't function with their own perceptions and knowledge base because it could lead to dangerous practices and animal mistreatment. Therefore, governmental bodies exist to make the rules and regulations clear for everybody.

Regulations are built on biosecurity and ethics. These rules will shift according to where you are situated. For example, in Western Australia, you need to register your animals and classify whether you are using them for commercial purposes. In New Zealand, the ethical concerns outlined in their regulations require you to provide your llamas with a constant supply of water and have a feeding setup that reduces waste. Furthermore, you must ensure that your llamas do not have access to materials they can mistakenly ingest that would cause them harm, like loose wires, building paper, electrical fittings, or shards of plastic. In the UK, you will need a CPH number and a HERD mark, and you will also need to register with the British Llama Society if you want to keep these animals legally. You will need an NLIS registration in Australia, the National Livestock Identification System.

You also have to consider your use of llamas in a broader sense because some places don't allow animals that llamas work the best with, like Arizona, which does not allow sheep to be kept. In America, different states have widely varying regulations, so you need to do your due diligence and conduct thorough research to ensure you are functioning within the barriers of the law. In Colorado and Georgia, you can own llamas and pets. In Idaho, the barriers to keeping llamas are minimal because you do not need any permit for domestic animals. Therefore, review what you can do in your region to determine whether llamas as guardian animals are the best option for you in the legal context you find yourself in.

Multispecies Grazing and When It Should Be Applied

Multispecies grazing is when you are keeping two or more different animals in the same pasture. They do not necessarily have to be grazing at the same time, but if they share the land in short intervals, it still counts as multispecies grazing. Using multispecies grazing can be good for the environment because the different habits of various species add to the area's biodiversity. You will most likely pair llamas with sheep or goats because they share a diet and protect each other from their common natural enemies.

Llamas' susceptibility to parasites requires you to use rotational grazing. The lifecycle of parasites, from eggs to fully grown, runs for a certain period. Rotational grazing is when you take your animals to different pastures at different times of the year so that your land can have time to recover and you can avoid the dangerous parts of the parasite lifecycle. Different regions have varying parasites, so you must research the common parasites in your area and the best rotation schedule to use so that your sheep, goats, or llamas don't get sick.

Horses and cows share parasites, while sheep and goats share parasites. Therefore, when applying multispecies grazing, you should pair sheep with cows and goats with horses. In the case of llamas, this kind of parasite management is not possible because they are guardian animals. You must monitor them to see if they are weak or fatigued because it could be a sign of parasite infection. If you feel your animals are sick, make sure to call the vet immediately because if you wait too long, it could end up being fatal.

As a protective guardian, llamas pair well with sheep. You can also use them for chickens, but they will not be as effective. They are even less useful for cows because one of their biggest benefits is their size, and cows are bigger than llamas. Furthermore, they would be unable to fight animals targeting cows. Introducing llamas into a complex system of multiple animals recreates the conditions of the natural world, and it is great for your plant life and the biodiversity of the ecosystem you inhabit. The use of llamas forces multispecies grazing because they are only effective if they spend a lot of time with the flock they watch over. Therefore, in addition to being one of the most captivating guard animals, llamas are an eco-friendly option.

Chapter 8: Training Your Llama

Now that you understand how llamas function as livestock guardians, you need some more details on how to train them so that you can maximize their effectiveness. Llamas are made for the role of guardians, but a few missteps can radically decrease their functionality. Therefore, you must be informed on exactly how to treat your llama so they can be seamlessly integrated into your flock and the daily routines of your farm or homestead.

Llamas can be kind, but they also have a terrible mean streak. Much of your training will be focused on managing their aggressive tendencies and mood swings. Llamas spit, bite, and kick when they are angry, which puts you, other people on your farm, and your animals in danger. Instituting the correct protocols and interventions can ensure that your llama stays calm and every other living thing on your property is safe, except for the predators.

Llamas require a lot of training.

From choosing the right llama, considering factors like size, age, gender, temperament, and personality, to finding ways to introduce your animal to their flock – all the steps of training a llama are covered. These practical tips will give you the knowledge you need to be an expert guardian llama trainer and get the most out of your animal. How well your llama performs solely depends on the work you put into training it. It may seem intimidating at first, but by understanding a few basic principles, you can be put on the fast track to becoming a pro.

Getting your llama on the same page as you requires mutual respect and understanding. You will dive deep into the psychology of your llama to set yourself up as the leader of its flock. Furthermore, you will maximize its protective and territorial qualities so that it can serve you well as a guardian. Llamas are already natural guards, so it does not take much to get them in tip-top shape for peak performance of their duties. You just have to lead your llama in the right direction toward its destiny as an elite guardian animal that will keep your livestock safe from many kinds of predators like foxes, dogs, and coyotes.

How to Choose a Llama

Choosing the right llama is the beginning of training. The right temperament, size, and personality will weed out the unsuitable candidates. If you were a basketball coach, you would choose the tallest, strongest, fastest, and most athletic people to be a part of your team if they all had no previous basketball experience. You would test their hand-eye coordination and their reflexes as well. Similarly, you should choose a llama that is best suited to its role as a guardian. This llama selection process will determine the level of difficulty you experience when training your llama for guard duty.

The age of a llama is very important when making your initial selection. Ideally, you want a male between the ages of 18 months and two years old. Once you have found a male that's the appropriate age, watch how he behaves. The llama should be accustomed to being handled by people because you will need to trim their toes, transport them, as well as sheer them. Ask the breeder to walk the llama around, and if possible, bring along a dog to see how the llama reacts to it. Llamas instinctively hate dogs, but some who grew up around domesticated dogs will not see them as an enemy, only targeting unfamiliar invaders.

Many llamas are kept as pets. This means the llama you want to purchase may have been brought up alone and raised on a bottle. These types of llamas are terrible as guardians because they bond more with humans than other animals. The pack behaviors of llamas, which make them such brilliant guardians, are learned in the first few months of spending time with their mother. Therefore, your llama must be on its mother's teat for at least six months before you buy it. Llamas hone their protective instincts by spending time with other llamas or flock animals. A great sign that a llama is not attached to people is if it exhibits minimal curiosity upon seeing you and acts aloof.

The male you select must be gelded – which means castrated. Gelded males are safer to work with because they will not challenge you as the leader. Moreover, they will not attempt to mount your other animals, which could cause injury. The llama must be gelded before it is bred because doing it afterward does not erase the threat to you and your animals. Testosterone-filled males who have not been castrated can become a huge problem because of their unpredictable tempers.

Observe how the llama moves and interacts. You want a level-headed llama that is not overly confrontational. Therefore, the llama you choose should not be screaming or spitting at people. Assertive llamas will attempt to chest-butt you or will stand their ground instead of moving out of the way when you approach it. When a llama shows these headstrong behaviors, it is best to keep looking. More signs that you are dealing with an alpha that will challenge your position is if the llama aggressively protects its food and refuses to allow people to clean its manure.

Using a reputable breeder is advisable because it is more likely that they will be honest and provide you with a strong and healthy llama. Breeders typically have to register with various regulatory bodies. Find out which organizations govern the industry in your region and seek out breeders who are aligned with them. If you can find a breeder strictly dedicated to raising guardian llamas, go with that option because the additional money you pay will be worth it. A guardian llama will set you back about $500 to $1,500.

If other animals are on the breeder's property, keenly observe how the llama interacts with them. A guard llama should be alert and curious about changes in its environment. They should be gentle with the animals around them. If you have pet dogs, ensure they are comfortable around them because dogs are natural enemies. Watch if the llama attempts to break the fence or get past it because you do not need an animal that is prone to escaping. Check if the llama is protective of young members of the flock. A guardian llama will either embed itself in the middle of the flock or walk up to higher ground to see better, so any of these behaviors is a great sign for an eligible candidate.

Fundamental Principles of Llama Training

The training for guardian llamas is not extensive because you are tapping into their natural behavioral patterns to meet your requirements. Like people, llamas are individuals. What works for one llama may not have the same impact on the next. Spending time with your animals allows you to get to know them. You'll learn what your llama likes, how it responds to different commands, and what causes it discomfort or stress. Llamas are prey animals, so they can be jumpy and easily startled. Be gentle and move deliberately so as not to frighten your animal. Take your llama out on walks, helping it become familiar with the property and comfortable with you as its feeder.

Llamas are quick learners, which can be a gift and a curse. You must be careful of the actions you take when training your llama. It will be very difficult to undo the impact if you inadvertently create a negative connotation to something, whether it is an object or an animal. Work with your llama every day because repetition will ingrain behavior, and you will learn your llama's personality better. Once you understand your llamas' individuality, you can adapt your training to be personalized to what it responds to.

Patience and calmness are the names of the game. Llamas are skittish and can be easily startled. You do not want to scare your llama into disobedience. Furthermore, rushing through training can cause you to overlook key details. Therefore, take your time each day and set reachable milestones. Your llama will only be as good as the training you are willing to put in. Luckily, the only areas you need to focus on are transporting your animal, integrating your llamas into the flock, and reducing aggressive behavior that can hurt people or your livestock. These sections of training can be easily addressed with the right information and minimal effort.

Step-by-Step Instructions for Building Protective Behaviors

1. Pair your llama with the right animals. Guardian llamas are best suited for sheep or goats. Their size advantage over predators does not apply to cows because they are bigger. Furthermore, they eat a similar diet to sheep and goats, which is important because in order for the llama to bond with a flock, it must spend time together. Llamas do not pair well with chickens because they often ignore predators that are dangerous to poultry, like weasels and skunks.

2. Make sure the llama bonds with your flock. To establish a bond or introduce your llama to your flock, you can place it in an adjacent enclosure. The bonding period takes about a few days to a couple of weeks.

3. Use only one llama. The problem with using multiple llamas is that they will become protective of each other instead of bonding with your livestock. One llama can cover 300 acres and can watch 300 sheep. If you have too much land or too many animals for one llama to be effective, you must make sure that your guardians do not meet.

Positive Reinforcement Techniques

Llamas, like many other mammals, are food-motivated. The best positive reinforcement you can adopt is using food as a manipulation tool. As a guardian llama, you may have to lead your animal into various sections of your farm. Therefore, you will need to train your llama to be comfortable getting led into different places. Using positive reinforcement when transporting your llama will ensure the process runs smoothly. Immediately reward your llama with a delicious, sweet treat, like apples or carrots, when it displays behavior that you like. This will create a positive connection with the behavior, which will cause the llama to keep repeating the desired action.

Positive reinforcement works best when coupled with a regular schedule. Feed and groom your llama at the same time every day. This routine reinforces desired behaviors with repetition. So, take your llama out into the pasture in the morning and bring him back in the late afternoon, giving him a treat when he enters the enclosure and leaves the enclosure. This way, you create positive reinforcement of its daily routine. Llamas can be stubborn, so getting them to enjoy being transported will help you in the long run because the animal will not resist being moved.

Integrating a Guardian Llama

Integrating your llama means getting him used to your unique farm environment. On your farm, you may have a variety of animals and other people that your llama will come into contact with. Your llama needs to be accustomed to all the interconnecting parts of your land, including other humans that work with the animals. Therefore, if people are interacting with your llama, they must spend time with him so that the llama can be well acquainted with the person they will regularly be in contact with. To reduce aggressive behavior, the llama must know that either you or anyone else they work with are a part of the flock. Try to limit human interaction because you want the llama to protect the animals and not form powerful bonds with people; otherwise, it will end up recognizing humans as the ones that need protection.

Llamas are extremely territorial, as they are herd animals that usually need to compete for resources. Its territorial nature can cause it to hurt members of your flock by kicking or biting. They can be especially violent toward other llamas, particularly if they are both males. You should only have one guardian on your property at a time to stop the llama-on-llama violence. To prevent the llama from attacking your flock, you must introduce them slowly. Keep the llama in an adjacent pen where it can see and interact with the flock without coming in direct contact with them. Your llama should be feeding at the same time as your flock so that it can make the connection that it is part of them. Gradually begin allowing them to interact. In the first few weeks, you must watch them closely so that you can intervene if there is any trouble.

Another way to prevent fights or attacks is to keep your sheep or goats in a big area with your llama. Confined spaces are breeding grounds for aggression. If your llama has enough open space, then the likelihood that it will attack your flock is significantly decreased. Llamas also have a strong sense of smell that guides their actions. You can use scent to your advantage when integrating your llama. Place hay or other objects with your sheep's scent in your llama's temporary pen. This will get your llama accustomed to the animal's smell, which will ease the integration process when you allow them to start spending time together.

Your llama needs to know where it is allowed to go and the borders where its territory ends. Walk your llama along the fence of your pastures and enclosures for your flock. Correct the negative behavior of trying to cross over the fence by yanking the lead and yelling a "No!" command. Introducing the llama to its space is just as important as introducing it to its flock. Furthermore, as a guardian, your llama must be familiar with its surroundings so that it can more easily spot danger.

Llamas and dogs are natural enemies, so if you have canines, you must either keep them separate from your llamas or introduce them so your llamas can accept your pet as part of its flock. Allow your dog and llama to meet under careful supervision so that you have control over both your animals. Correct the behavior if any of them show signs of aggression. You may need to repeat this process multiple times because you are attempting to override a natural fear, which could take a lot of time and effort.

Practical Solutions for Dealing with the Most Common Challenges during Llama Training

Some of the main challenges that can arise when training your llamas are aggressive behavior, like challenging you or spitting, distress, being easily startled, and forgetting some of the training you have already done. All these factors can be remedied with immediate intervention. Llamas, like dogs, can be trained with a combination of hand signs and voice commands. The best way to get the llama to internalize these commands is by using food as a reward. Positive reinforcement, just like food rewards, works with getting your llama to display behavior that you like, but it can be a bit tricky when you want to break negative behavior.

The challenge of aggressive behaviors from a llama starts with the selection process when you choose a llama according to its temperament. However, llamas that seem level-headed at first can become more aggressive due to an environmental change or a traumatic event. When fighting aggressive actions like spitting, screaming, or chest-butting, the main principle is to address the behavior immediately when it occurs. You cannot let this negativity slide even once because it will let the llama think that the way it is acting is fine. Llamas learn by hand gestures and commands, so you can put your hand up and sternly yell "no." Another option you can apply is firmly pushing against its chest while shouting "no." Some people use a water gun or a spray bottle as a repellant. You must repeat your "no" command every time your animal is aggressive so that it will be constantly reminded that the behavior is unacceptable.

Sometimes, it is not anger that causes undesirable behavior; it is distress. You can gently hold your llama to calm it down when it is feeling shocked, anxious, or stressed. Controlling stress responses is all about being mindful and not scaring the nervous animal. Remember that once a negative connection to an object is established, it will be nearly impossible to break. Therefore, during key times, like transporting and feeding, you must be extremely careful and prevent your llama from getting startled in these crucial moments.

Stubbornness is another behavioral trait that is common among llamas. Everything may be good in one moment, and in the next minute, you get zero percent cooperation. This is the perfect scenario to exercise patience. First, investigate what went wrong. The llama may have spotted something that made it uncomfortable, like a dog or a predator. If you have assessed that there is no perceived danger for the llama, you may need to use a little force. By no means does that equate to hurting your animal in any way. Training methods that cause harm are completely off the table because ethically treating your guardians is essential. Simply yank on its lead and shout your chosen command, such as "Yep!" or "Go!" At times, you may need your llama to move immediately, so getting it accustomed to such a command is crucial.

The seasons on your farm may necessitate the adoption of differing routines depending on the time of the year. Your llama may forget what it was taught in the previous season, so you may need to retrain it. Repetition is what allows the llama to internalize schedules and commands, so when the repetition is broken, it can easily lose what it had learned before. This is why training llamas requires you to have a lot of patience. You may find yourself having to retrain your animal multiple times a year, depending on how your schedule changes. Make time for an adjustment period when you are making shifts. Radical changes can cause a lot of stress for your llama, so be sure to respond according to its emotional condition.

Llamas are so well-suited for the position of guardian animals that they do not require much training. As long as you maintain a healthy physical and mental state for your animal, there should be no problems. You can make some corrections for aggressive behavior and for introducing the llama to your flock, but for the most part, your llama will automatically jump into its duties.

Chapter 9: Guard Donkeys

To make one thing clear, not all donkeys are like Eeyore. They are social and fun creatures, albeit a little stubborn. They are also guard animals that can protect your livestock. These nice-looking animals have an aggressive nature that usually comes out when they are threatened, making them the ideal guardians. Once your livestock see how donkeys protect them from harm, they will gravitate around them to feel safe.

This chapter covers the unique characteristics of donkeys, their protective nature, what makes them great guard animals, and the advantages of donkey guards.

Donkey as a guardian.
NasserHalaweh, CC BY-SA 4.0 <https://creativecommons.org/licenses/by-sa/4.0>, via
Wikimedia Commons.
<https://commons.wikimedia.org/wiki/File:Equidae_Equus_africanus_asinus.jpg>

Donkeys' Unique Characteristics

Every animal is unique in its own way. Most people would tell you dozens of facts about cats and dogs, but if you ask them about donkeys, they won't have much to say. Donkeys are extremely underappreciated, but they are amazing animals with some interesting qualities that make them fun to raise.

They Have Big Ears for a Reason

Have you ever wondered why donkeys have large ears? Many of them evolved in dry conditions, like in Asia and Africa. Their large ears expel the heat inside their body to regulate their temperature so they can withstand the hot weather and remain cool during the summer. They also boost their hearing capabilities so they can pick up on mating calls and when a predator is approaching.

Some Donkeys Are Very Small

Have you ever seen tiny donkeys? They are ridiculously cute, right? Some types of donkeys are really small. They are about three feet tall and are very common in Sardinia and Sicily. Interestingly, the shortest donkey in the world is 25.29 inches high. Although many small animals breed, the donkey's small size is natural.

Donkeys Are Stubborn

Donkeys are very stubborn animals. Sometimes, they plant their feet firmly on the ground and refuse to move from their place. Even if you try to pull them, they will not budge. Understand that they are not being jerks or dumb, as some people assume. In fact, they are very clever animals. If donkeys feel threatened or in danger, they will stay put to give themselves time to think if it's safe to keep moving. They are pretty smart, aren't they? This is unlike horses, which run away when they are afraid.

They Are Very Social Creatures

You will rarely see a donkey alone in the wild. Donkeys are social creatures that like to be in groups. It is in their nature since they usually live in herds and form strong and lifelong friendships with other animals. Two donkeys can form a close bond, which is called a "pair bond." They become very attached to each other, and they would suffer from stress, anxiety, loss of appetite, and discomfort if they were separated. So, it is better to adopt two donkeys to keep their spirits up and keep each other company.

They Are Hairy Creatures

Some donkey breeds are very hairy and look very pretty. One of the most popular types is the Poitou donkey, which originated in the French city of Poitou. These donkeys have long and thick hair and are usually very expensive.

Unique Voice

Donkeys have a distinct sound called "braying." Unlike zebras, horses, and other animals, donkeys can make this sound while breathing. They make the "hee" sound while inhaling and the "haw" sound while exhaling. They use their unique vocalization to communicate and connect with other donkeys. They are usually very loud, so they can call on other animals from a distance.

They use their unique vocalization to protect other animals by braying loudly to alert them of any potential dangers.

More Than Work Animals

Many people treat donkeys as just work animals. They depend on them for transportation or to guard their livestock because of their endurance, strength, and ability to adapt to tough conditions. However, there is more to donkeys than what meets the eye. They are sociable, intelligent, affectionate, loyal, and can form deep relationships with their owners. Your donkey will feel like family, not just a guard animal.

Adapt to Dry Conditions

Donkeys can survive and easily adapt to harsh and dry weather conditions. Their body can conserve water as their kidneys extract water from different organs, so they can stay hydrated for long periods of time.

They Are Light Sleepers

The next time you see a sleeping donkey, tiptoe around them, or you will disrupt their slumber. Donkeys are light sleepers and only nap for a couple of hours every night. However, they are usually cautious and alert during this time, so even while resting, they remain vigilant. They sleep while standing up but lie on their back or side when they need to rest.

Donkeys Are Therapy Animals

It isn't just cats and dogs that can be therapy animals; donkeys can be, too, thanks to their intuitive and gentle nature. They offer companionship and support to mental health patients and people with disabilities. They can sense when you are going through hard times or experiencing depression or anxiety.

Donkeys Are Fast Learners

Donkeys are intelligent, resilient, and pragmatic animals with a great memory. They are fast learners and can understand instructions quickly.

Male vs. Female

Male donkeys are called "Jacks," and female ones are called "Jennies," and each has their own fighting style. Jacks kick their enemies with their front feet, while Jennies use their hind legs.

Hormones

Female donkeys experience hormonal changes every month, which can lead to behavioral issues. They can either be very angry or super friendly. If you plan to raise Jenny, you should be understanding and patient during this time. You should also log her cycle dates on your calendar to be prepared every month. However, some female donkeys don't show any sign of behavioral changes during their cycle, and their temperament remains the same.

Protective Nature

It is a donkey's nature to protect other animals, like goats and sheep, from roaming dogs, coyotes, and other predators. They are territorial and will attack any stranger that invades their space. Donkeys are also very strong and are able to handle themselves against many strong animals. However, they are friendly and affectionate to human beings.

Female donkeys (or Jennies) are more protective than their male partners since their motherly instincts drive them to protect weaker and smaller animals and keep them safe.

Some people believe donkeys don't deliberately protect farm animals – they protect themselves and their territory. Whatever their intentions are, donkeys will keep your animals from harm.

How Donkeys Protect Their Flock

Donkeys don't always come to mind when people think of guard animals. In fact, you are probably wondering how a donkey can protect livestock. First of all, the donkeys should be present with the herd at all times. You can't just keep them with the animals for a few hours each day because you never know when a predator will attack. Donkeys also need to spend more time with the farm animals so they can bond together. This won't be hard for these social creatures who crave connections with other animals.

Donkeys rely on their unique hearing capabilities, which can make them detect noises from a distance, and their strong sight allows them to spot a predator from afar and take the necessary precautions before they attack. Sheep and other farm animals are very smart. They will quickly notice that the donkeys are their friends and allies and will seek their protection when threatened.

Donkeys scare their predators by letting out very loud brays and chasing after them to drive them away. This tactic should also get your attention and alert you that something isn't right so you can go and check on your animals. In most cases, you won't need to interfere, as the donkeys will confront the predator by themselves.

However, not all canines will retreat right away. In this case, the donkey will attack them by kicking them with their front feet, injuring or killing the predator. Male donkeys may also bite.

A donkey's strong herding instincts, aggressiveness, and natural dislike for predators make them ideal guard animals.

Advantages of Using Guardian Donkeys

Unfortunately, donkeys aren't as popular as guard animals as dogs are. People think they aren't strong or clever enough to protect their livestock. However, there are many reasons to consider guard donkeys.

Donkeys Are Always Alert

Thanks to their big ears and peripheral vision, donkeys are always aware of their surroundings and alert to any impending danger. Even when asleep, they are still alert and able to pick up any strange noise close or far away.

Territorial Behavior

Donkeys are very territorial, and their protective instinct results from this behavior. In other words, they don't protect the herd but their territory. If they feel someone invading their space, they will attack immediately.

Unlike dogs, donkeys don't patrol their area. They are already alert and can sense danger without moving.

Compatibility with Livestock Animals

Donkeys are compatible with sheep, and they bond easily with one another. Most donkeys happily protect sheep from predators. However, you should introduce both animals to each other early on and raise them together. Even if they didn't grow up in the same place, you can still teach your donkey to protect the herd by letting them live next to each other for two weeks. Significantly, the donkey and the sheep are compatible to avoid conflict. They won't protect an animal they can't tolerate. They are also compatible with horses, alpacas, llamas, goats, pigs, and other donkeys.

However, donkeys aren't compatible with farm dogs or any type of canine, so you should be very careful when introducing them to each other.

Guard Donkeys Aren't Expensive

Inexperienced guard donkeys aren't expensive, but you must train them. This will be easy since donkeys are clever and follow commands. Some won't even require training – you just let them socialize with your flock, and they will follow their protective instinct. You can buy a well-trained donkey, but they are more expensive.

Better Than Dogs

Training dogs takes more time and effort than training donkeys. They may also attack your livestock instead of protecting them. Guard dogs tend to bark a lot, unlike donkeys, who are fairly quiet unless they feel threatened. Most farm owners prefer to live in a quiet environment, and the barking can be off-putting. Dogs are also considered predators and won't be able to relate to farm animals. Donkeys, on the other hand, can relate to farm animals since they are both prey.

Donkeys Are Independent

Guard donkeys are very independent; they only need shelter, food, and water. They don't require constant care or attention. They also won't need expensive vet care since they aren't prone to injuries.

Protect against All Animals

Guard donkeys will protect almost all livestock, such as goats, sheep, and even chickens. Unlike with guard dogs, you won't have to worry about your donkeys attacking or eating any of your livestock.

They Live a Long Life

Your guard donkey will be with you for thirty years or more. Unlike other guard animals, donkeys have a long lifespan. They are also cost-effective since you won't need to buy a new one every ten years.

No Separate Accommodations

Being a guard, donkeys won't attack your animals; you can keep them in the same pasture with your livestock. This is another advantage they have over dogs, which require their own accommodations.

They Are Large

Most donkeys are large in size (between 300 and 500 pounds), so they will be able to handle different predators like foxes and coyotes. Stick to large donkeys and avoid miniature ones. They are nice to look at, but they won't protect your livestock from predators.

Their Braying Is Useful

Unlike dogs, a donkey won't bray all night or expect you to come to their aid. However, their very loud "hee-haw" sound is a clear sign that a predator is nearby.

Unfriendly to Canines

Donkeys don't get along with dogs or any type of canine, so they will be on guard if they see one approaching. Some donkeys don't have a problem with pet dogs, but many don't appreciate their presence.

They Are Ready for a Fight

Donkeys will never run away from a fight – they will stand their ground. They don't scare easily or get nervous in unfamiliar situations. They are curious, confident, and courageous in the face of danger. In fact, coyotes, foxes, and other small predators avoid confrontations with donkeys at all costs because they know they are tough fighters.

Calm Temperament

Donkeys are very calm animals and won't pose any threat to you, your family, neighbors, or other animals.

Real-Life Story

Amanda and her husband, Taylor, are fond of cattle ranching. They love their animals and consider them family. Standing among their cows, walking around, and braying are her beloved donkeys. Amanda said her husband didn't want dogs or horses on their ranch. She decided to prank him and get donkeys instead. To her surprise, Taylor came home one day with three donkeys. She realized that her husband had always loved donkeys and had plans to add a few to his ranch for a long time.

Amanda immediately fell in love with the donkeys because they were very affectionate and friendly. However, they weren't pets and should be kept with the cattle to guard them. Amanda said raising the donkeys was easy, and they got along very well with their cows. They followed the herd around and ate and drank with them. The donkeys protected the cows and their calves from roaming dogs and coyotes. Amanda and her husband agree that donkeys are the perfect guard animals. They have been with them for seven years and have only lost one cow to coyotes.

Amanda and Taylor entrust their cattle with the donkeys all day while they are at work. Every day, they appreciate their little helpers more and more.

Caring for a Guardian Donkey

If you want your donkeys to stay healthy and live with you for a long time, you should take good care of them and pay attention to their feeding, shelter, and health.

Feed and Care

You should leave fresh water for your donkeys all day. Check it throughout the day and refill it when needed. One donkey requires 10 to 25 liters every day. You should add mineralized salt to their diet. Read the ingredients on the feed packaging or check with the district agriculturalist to find which minerals are missing from their food. To protect your animals from mineral or vitamin deficiency, find ways to add them to their diet, like giving them supplements.

During the winter, feed your donkeys high-quality hay. Avoid legume hay since it is high in protein. Brome grass and meadow grass are your best options. Feed pregnant and nursing donkeys 50 percent alfalfa and 50 percent Timothy hay. Feed male guardian donkeys grains and give them supplements to increase their energy levels. Avoid supplements made for poultry, pigs, or cattle, as they can be toxic to them.

Health

Donkeys require regular deworming and vaccinations. Deworm them three to six times every year. You can do this yourself using a paste warmer or get a professional to do it for you. If you suspect your donkey has parasites, call your vet right away to get them the proper vaccinations.

Proper hoof care is necessary for guard donkeys. Trim and clear out their feet every month. If you neglect hoof care, they will grow to a very large size and cripple your animal.

Donkeys also require dental care, so make sure to get their teeth checked twice a year.

Shelter

Donkeys prefer warm climates, but they can adapt to the cold if they are provided with enough food and a warm and safe shelter. Donkeys can't stand the rain since their coat isn't waterproof, so they are left cold and wet, which can lead to various diseases like bronchitis and pneumonia. Make sure their shelter protects them from the rain. You should also remove the snow from their coat during the winter. Keep your donkeys in the barn during the winter and only let them out on warm days. In the warm weather, they will only need an open-front shed bedded with dry straw.

Pasture

Let your donkeys graze coarser pastures, but avoid lush ones since they can increase their weight and cause other serious health issues. Allocate your donkeys one acre of pasture every month.

Donkeys are intelligent and strong animals that can protect your cattle against any threat. They have unique skills that can make them sense their enemies from a distance. Although there are different guard animals, donkeys have many advantages that make them a great choice.

Mankind has been depending on donkeys for centuries. They have used them for transportation, to guard their animals, and even to eat their meat and drink their milk. These faithful companions make life much easier, and they never ask for anything in return. Make sure to give them a loving home and a warm shelter during the cold. Take care of their health and protect them just like they protect you and your livestock.

There are still more things to discuss about donkeys, like how to choose the right one for you and how to train them. Head to the next chapter to find out all this information and more.

Chapter 10: Choosing and Training Your Donkey

Now that you know everything about guard donkeys, you are probably considering bringing one home to protect your livestock. However, you can't just choose any donkey. You have to go through a process to find the right one for you. Afterward, you will need to train them so they can become the best guard animal there is. Understandably, this is easier said than done.

This chapter provides expert tips and practical techniques for the donkey's selection and training process.

Training guardian donkey.

Selecting Guard Donkeys

There are a few things to consider before choosing a guard donkey, like age, breed, temperament, and gender. You don't want to bring a donkey home only to realize they aren't a good fit. Follow these tips to select the ideal guard donkey for you.

Age

Choose a donkey that is – at most – three years old because sheep and other cattle get along better with younger animals. Remember, the older the donkey, the harder the introduction will be. They will also be playful, and they will run around with the sheep and bond together. Some people prefer donkeys that are six months or younger to raise them with the herd. However, an animal that age won't be a good guardian. Older cattle may also bully them or play roughly with them.

Gender

Jenny and her foal (young donkey) are ideal for guarding donkeys. Females have motherly instincts and are natural protectors. However, the Jenny will be enough if you don't want more than one donkey. Geldings have also become very popular in the last few years because of their calm temperament. Avoid intact Jacks (unneutered males) because of their aggressive nature and tendency to be violent with people and cattle. Pregnant Jennies aren't a good idea since they will focus more on their newborns and ignore the livestock. If your flock is attacked, they will only defend their newborns with no concern for the safety of the other animals.

Size

Stick with large or normal-sized donkeys. However, some people don't prefer very large donkeys as they are hard to handle. There is no denying that they will scare off predators, but you should consider if it is worth the hassle. Choose a size of about 44 inches.

Avoid miniature ones because they are too small and don't have the physical strength to guard a flock. A small donkey won't be intimidating or scare predators. Your herd may also be attacked by a pack of animals, so you need a big enough donkey to fight them off and scare them away. A miniature donkey will still fight them because it's their nature, but they will get seriously injured or even killed.

Breed

All donkey breeds, like Mammoth, Australian, and Irish, can be guard animals. Just make sure you choose a healthy and sturdy one.

Temperament

Avoid donkeys with an aggressive nature, as they aren't easy to handle. Remember – you want a donkey that will protect your animals, not one that is easily agitated and can hurt them. Choose an even-tempered donkey that you feel safe leaving your flock with all day.

Flock Size

Donkeys can only guard a small flock of about 100 animals or less if they are scattered across the land. However, a donkey can guard about 200 sheep if your flock grazes on one pasture. So, you should consider getting more than one donkey if you have a large flock.

Other things to look out for:

- Good attitude
- Straight legs
- Good conformation

Where to Find a Guard Donkey

- "Adopt, don't shop" applies to guard animals as well. You will find many rescue donkeys in specialized organizations. They are usually tested with goats and sheep to assess their guardian skills. You can return the donkey to them if things don't work out. However, if things do work out, you will be giving a worthy donkey a loving home - and a job!
- You can also find donkeys in livestock auctions or a mule and donkey breed association.

Training Guard Donkeys

After choosing your donkey and bringing it home, you should start training it. You can get an experienced guard donkey, but they are usually more expensive. If you want to save money, learn to train your animal yourself. Luckily, this will be easy since donkeys are very clever animals.

Basic Principles of Training Guard Donkeys

The main principles of training animals are behavior correction and positive reinforcement. Your donkey's negative behavior should be corrected right away so they understand that there are consequences to their actions. Positive behavior should also be rewarded.

Positive reinforcement is one of the most powerful training techniques, as it can shape or change your donkey's behavior. It usually involves giving rewards like praise, treats, or anything your donkey enjoys. They will most likely repeat the positive behavior in order to keep getting the rewards.

Training your donkey or any animal requires a lot of patience. You are teaching your donkey to adopt new behaviors, but they are bound to make mistakes and return to their old ways, so you need to be patient with them. Getting your donkey to repeat the same behavior isn't always easy, even with them being clever. You may sometimes feel bored or frustrated, but you must exert self-control. You lose control when you lose your temper, which confuses the donkey. Always remain firm and calm, and you will see real results in no time.

Obedience Training Techniques

Instructions:

1. If your donkey is hyper or acting out, training them will be very difficult. They won't be able to focus and learn something new. So, before you start, let them do a couple of exercises to get into a better and calmer state.

2. Remove all distractions. Teaching an animal something new is already tricky, so try to eliminate anything that can affect their concentration. It is better to train them in a low-key environment.

3. Make the technique fun. It should feel like playing, not work. Enjoy your time with your animals, and take the opportunity to know them better.

4. Decide on a reward system and make sure it is something your donkey responds well to, like treats or praise.

5. Reward them with treats or praise when they follow your commands. If they don't obey, be patient, and keep trying... they will eventually get it.

6. Take a break whenever you start to feel tired or frustrated. The break can be a couple of hours or a few days. Be patient with yourself and your donkey.

Stopping Bad Behavior

- Don't punish your donkey because many animals don't understand the concept of punishment. It can do more damage than good.

- When your donkey misbehaves, walk away from them. Donkeys that are bonded with their owners will not want to lose the attention you give them. They will try to rectify their behavior to regain your bond and trust.

- You can also withhold something they like from them, just like you do with a child. For instance, if they bite or fight with the cattle during mealtime, withhold their food from them until they correct their behavior. When they do, reward them with positive reinforcement, like treats or food.

Positive Reinforcement vs. Treats

Treats are effective, but you can't reward your donkey for every good behavior. This will teach them to act favorably only to get something in return. They will come to expect treats each time and may act out if you don't give them any. So, use treats only on occasions and when you want to correct very bad behavior, like biting. Stick with positive reinforcement instead, like petting, scratching, praise, or positive talk.

Train Your Donkey in a Small Area

Training your donkey in a small area will put you in control of the situation. This will prevent them from running or hiding. It will make things easier since you will be able to limit distractions, so your donkey will only focus on you.

More Training Tips

- Your donkey will learn something new every time you interact with it.

- Donkeys don't understand the difference between good and bad behavior. Their behavior is either effective or ineffective for them.

- Donkeys learn activities related to their natural behavior faster. It takes them longer to learn things that are unnatural to them and detached from their nature, like traveling in a trailer, holding their feet up for hoof cleaning, pulling a cart, and being ridden or driven.
- Start with teaching your donkey relatable behavior, like getting food or walking next to you.
- Using positive reinforcement when teaching your donkey natural behavior will make it easier to teach them unnatural ones.
- Never discipline your donkey or punish them; they will get angry and stop cooperating.
- Don't use sticks when interacting with them; instead, use carrots.
- When they don't follow directions, switch to another familiar activity.

Patrolling

Donkeys don't patrol the area – they usually look for any threat within their flock. Although this behavior isn't in their nature, you can train them to patrol. For instance, you can take them on walks to discover the area so they get used to it and know it's safe. You can also ask someone to make a sound in the distance and take your donkey to investigate. This will teach them to patrol the area when they hear a strange sound. Use positive reinforcement or teats to reward them for patrolling.

Alertness and Responding to Threats

Although donkeys are naturally alert, you can still train them to boost this behavior. Reward your donkey whenever it brays or exhibits any guarding behavior at any sign of danger. This teaches your donkey to be more alert to predators and threats.

You can use the same technique to train your donkey to respond to threats. As donkeys are territorial, they will naturally be protective over the livestock. Use treats or positive reinforcement whenever your donkey responds to threats; this encourages that behavior more.

Socializing Your Donkey with Your Livestock

Introduce your donkey to your cattle right away to initiate the bonding process. The donkey will only exhibit protective behavior when they feel the sheep is part of their flock. Raise Jenny and her foal with the cattle. Weaned foals can be left alone with the cattle.

There are other steps you should take to help your donkey socialize with your livestock. Understand that this won't always be a simple process. Some donkeys are prone to territorial behavior, and it can take them a while to accept other animals. Depending on the donkey, the introduction can take an hour or a few days. However, the bonding process takes about five weeks. Make sure to have someone with you while making the introduction in case the donkey or the herd reacts unfavorably.

Instructions:

1. Let the donkey stay in the same barn as your livestock for two weeks, but separate them with a fence.

2. After the two weeks, bring them to the same pasture and put a fence between them. Give them time and space to sniff each other so they can become familiar with one another's scent, but keep an eye on them.

3. You can lead the donkey around the flock with a rope so they can smell one another. Repeat this until they accept and trust each other.

4. Feed the donkey with the cattle so they can bond. The donkey will also feel that it's a member of the herd.

5. Once they accept each other, allow the donkey to run around on the pasture. In time, the cattle will seek the donkey whenever threatened.

Signs the Donkey Isn't Reacting Favorably

- Lunging
- Biting
- Ears pinned
- Head down
- Chasing
- Ears pulled back
- Flared nostrils
- Whiteness around the eyes.
- Exposed teeth

Expect some minor fights, but this isn't any cause for alarm. They are just getting to know each other. Only intervene if they start to act aggressively. However, don't stand between them. Create a distraction so they would pay attention to something else. Don't introduce them to each other right after the fight. Give them more time to get used to each other's scent, then try again. Remember – donkeys are stubborn, so things may not go as you planned. Don't let this discourage you, though. Since donkeys are also social creatures and love bonding with other animals, they will eventually get along. Just give them time.

Get Your Donkey to Trust You

It isn't enough that your donkey trusts the cattle – they should trust you, too. Creating a strong bond between you and your guard animal will make training much easier.

Instructions:

1. Visit their area a few times a day.

2. Stand there and don't approach them.

3. If they approach you, pet them or give them a treat.

4. Try standing a little close to them while monitoring their body language.

5. Retreat if they look uncomfortable.

6. If they come to you again, give them another treat.

7. Repeat this every day until they learn to trust you.

Introducing Donkeys to Predators

You can't bring a coyote to your farm and let it hang around your livestock. Your safest option is a dog.

Instructions:

1. Put your dog on a leash and bring it to the donkey to make the introduction. The donkey should be tied or held.

2. Keep a distance between them, but make sure they can still see each other.

3. Approach the donkey slowly with the dog.

4. Use commands like stay, wait, stop, sit, come, etc., to control your dog if the donkey acts out.

5. If the dog seems interested or curious about the donkey, stop approaching and give your dog a treat.

6. Take another step while monitoring your dog. If they remain calm, give them another treat. If the dog or the donkey seems scared, calm them down.

7. When both are calm, approach gradually while keeping an eye on both of them.

8. Once you get close to the donkey, allow them to sniff each other. Do this for a couple of seconds, then retreat and give each animal a treat.

9. Repeat this process until they become familiar with each other. Make sure to keep both of them leashed so you can remain in control and protect them from harm.

10. Once they seem comfortable, remove the leash and let them interact.

Safety Tips

• Monitor their interactions at all times because your dog may bite the donkey or get too close to it, making the donkey uncomfortable.

• If the donkey gets irritated or agitated, it might kick the dog and seriously injure it. Keeping an eye on both of them will guarantee everyone's safety.

Safety Measures

You need to keep yourself safe when training your donkey. Donkeys may not react favorably to training, so you should be prepared for anything.

Handling a Donkey

• When approaching a donkey, speak in a calm voice to get their attention.

• Don't approach them head-on or from behind, or they may consider you a threat. Instead, approach by their shoulder.

• Gently touch their neck and keep touching them so they are aware of your location.

• Stand by their side to protect yourself from getting kicked.

- Keep it on a leash if it is prone to kicking or aggression.

Safety Precautions

- Monitor your donkey's body language during training sessions and take a few steps back whenever they seem uncomfortable. Remember, donkeys are clever animals that may kick or bite you from behind to take you by surprise.
- Training sessions should be in spacious areas so you can easily get away whenever they show any sign of aggression.
- Avoid sudden moves or raising your voice. Speak calmly and use a gentle touch to prevent agitating them.
- Aggression is a sign of fear, so try to be understanding.

Common Challenges

Charging

When a donkey is angry or threatened, it can charge at you or another animal. If your donkey is about to attack, walk away from it. Stand across the fence until it calms down and realizes that you are not a threat. Luckily, charging isn't a common behavior among donkeys.

Kicking

Donkeys have very strong kicks, which is one of the reasons why they are great guard animals. If they kick you or an animal of your flock, they can cause serious injuries. Kicking is a sign of aggression, so treat the situation with caution.

- Leave them alone for a few hours until they calm down.
- Find what triggered their aggression and deal with it. For instance, if they are afraid of another animal or person, separate them from each other.

Biting:

Biting is another sign of either aggression or fear. They will either bite you or the flock they are supposed to protect. For instance, they may bite and toss smaller animals or birds like baby pigs, chickens, or ducks. They may also bite you if you are making them do something against their will. You can change this behavior by using the tips in this chapter.

Snorting

Snorting is a sign that your donkey doesn't want to be bothered today. They are either mad about something or want to be left alone. Give them some space, then check on them.

Flaring Nostrils, Pinned-Back Ears, and Pawing at the Ground

These are all signs of aggression. Your donkey may act this way because it feels threatened. Before you react, make sure that there aren't any predators nearby. Try to find what triggers this behavior. If it is afraid of another animal, separate them, wait until they calm down, and make gradual introductions again.

More often than not, your donkey's aggression is a sign of fear. You must eliminate whatever triggers these emotions and make them feel safe. When you become familiar with their behavior, you will know whether they are acting out or scared. Use any of the tips in this chapter to rectify their behavior. Remember to always be calm and patient with them and treat them like children.

Conclusion

Now that you've explored the details of livestock guardians on all levels, you are ready to begin your journey with these protective beasts. Do not rush to make any decisions. Take your time and consider all the aspects of your land and the animals you are raising. The right guardian can keep your livestock safe, but the wrong choice can increase your stress. Taking on a livestock guardian as part of your farm family is not a small decision to make. Think long and hard about whether you are ready for the responsibility and how introducing this animal to your livestock will help you. Do not wear rose-tinted goggles when assessing your farm, and be completely honest with yourself so that you can make the most beneficial decision for you and your livestock.

Take the calculated steps to first check which predators are plaguing your land and how they behave. Then, consider which guardian species or breed best aligns with your individualized context. Whether you choose dogs, donkeys, or llamas, it will all depend on your specific contextual environment. Consider how big the predators you are facing are and how the guardian responds to them. When you have replayed every possible scenario in your mind multiple times and looked at your farm through a microscope, you will be ready to take the next step and identify a guardian animal.

You are crafting an artificial ecosystem with the animals you introduce to your farm. The animals you bring together must create a seamless dance of cooperation. There may be hiccups and conflicts along the way, but as the head of the ecosystem you created, you can intervene when necessary. You are the choir conductor who creates the melody of your farm. Carefully analyze the fine details of your biodiverse approach to farming so that you can easily respond to any challenge or required change.

Be mindful of your environmental impact. The earth, which includes your farm, provides you with literally everything. Leave the planet in a better condition than you found it so that future generations can thrive. Using livestock guardians is already a step in the right direction because it is superior to more detrimental methods of predator control, like poisoning. It is not about humankind conquering nature but rather about building an environment where both can symbiotically coexist.

Raising a guardian animal is a big responsibility. There are countless care considerations you must make, like nutrition, parasites, and medical concerns. Your guardian is doing an important job, so the least you can do is make sure that their well-being is optimal. You must care for your guardians so they can help you care for your other animals in a beautiful balancing act of nature. Predators can be the downfall of a farm, so a good guardian is intrinsically tied to your success as a farmer. Therefore, you must do all you can to ensure they are emotionally, physically, and mentally at ease.

Here's another book by Dion Rosser that you might like

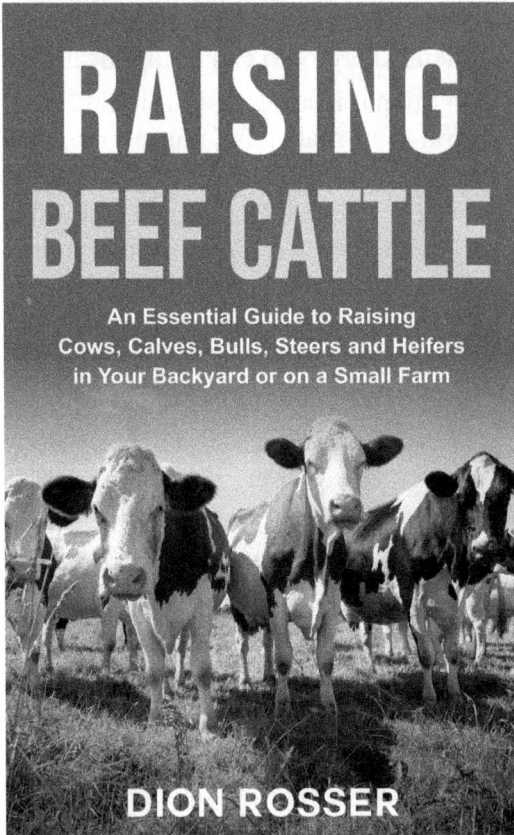

RAISING
BEEF CATTLE

An Essential Guide to Raising
Cows, Calves, Bulls, Steers and Heifers
in Your Backyard or on a Small Farm

DION ROSSER

References

(2021). Unfccc.int. https://unfccc.int/sites/default/files/resource/SB2021_01.pdf

(N.d.). Homesteadingtoday.com. https://www.homesteadingtoday.com/threads/lgd-behavior-problems-how-to-resolve.386665/

(N.d.). Tamu.Edu. https://sanangelo.tamu.edu/files/2020/06/LGD-Puppy-2020-Final.pdf

(N.d.). Wur.Nl. https://edepot.wur.nl/283765

(N.d.-a). A-z-animals.com. https://a-z-animals.com/blog/10-incredible-donkey-facts/

(N.d.-b). Pethelpful.com. https://pethelpful.com/farm-pets/Twelve-Fascinating-Things-You-Never-Knew-about-Donkeys

Ali, Y. (2021, September 16). Livestock Guardian Breeds: Get to Know These Working Group Members. American Kennel Club. https://www.akc.org/expert-advice/dog-breeds/get-to-know-the-livestock-guardian-dog-breeds/

Animal predatory behavior. (n.d.). Psychology Wiki; Fandom, Inc. https://psychology.fandom.com/wiki/Animal_predatory_behavior

Barnes, A. (2021, March 23). Creating an enriching life for llamas. The Open Sanctuary Project; The Open Sanctuary Project, Inc. https://opensanctuary.org/creating-an-enriching-life-for-llamas/

Barnes, A., & Hess, T. (2018, April 12). Hello camelid companion! The new llama arrival guide. The Open Sanctuary Project; The Open Sanctuary Project, Inc. https://opensanctuary.org/the-new-llama-arrival-guide/

Barth, B. (2017, September 14). How to Choose a Livestock Guard Dog. Modern Farmer. https://modernfarmer.com/2017/09/choose-livestock-guard-dog/

Bennett, S. (2022, July 12). How to train A donkey: A simple guide. Farm & Animals. https://farmandanimals.com/how-to-train-a-donkey/

Boi, L. (n.d.). Three lessons from training llamas. Utah.edu. from https://accelerate.uofuhealth.utah.edu/improvement/three-lessons-from-training-llamas

Bryan, M. (2021, September 21). Donkey facts. Facts.net. https://facts.net/donkey-facts/

Bukowski, J. A., & Aiello, S. (2023, November 14). Routine Health Care of Dogs. Merck Veterinary Manual. https://www.merckvetmanual.com/dog-owners/routine-care-and-breeding-of-dogs/routine-health-care-of-dogs

Caring for Guardian Donkeys Farmers are increasingly turning to non-lethal techniques for predation management. Donkeys have become a popular protector of sheep that can perform very well under certain conditions. Following are some guidelines on management and caring for guardian donkeys to maximize the animal's capacity in regards to flock protection. (n.d.). Ontariosheep.org. https://www.ontariosheep.org/uploads/userfiles/files/-%20Caring%20for%20Guard%20Donkeys-.pdf

Department of Jobs, Precincts and Regions. (2023, November 3). Guard dogs. Agriculture Victoria. https://agriculture.vic.gov.au/livestock-and-animals/animal-welfare-victoria/dogs/guard-dogs

Differences between llamas & alpacas. (2022, March 23). Animal safari. https://animalsafari.com/whats-the-difference-between-a-llama-and-an-alpaca/

DipCbst, A. F. C. (2023, May 17). 11 Tips for Training Livestock Guardian Dogs. PetHelpful. https://pethelpful.com/dogs/Tips-for-Training-Livestock-Guardian-Dogs

Dohner, J. (2014, April 24). Selecting a guard llama. Mother Earth News – The Original Guide To Living Wisely; Mother Earth News. https://www.motherearthnews.com/homesteading-and-livestock/selecting-a-guard-llama-zbcz1404/

Dohner, J. (2014, February 25). More questions to ask yourself before selecting a livestock guard dog. Mother Earth News – The Original Guide To Living Wisely; Mother Earth News. https://www.motherearthnews.com/homesteading-and-livestock/selecting-a-livestock-guard-dog-zbcz1402/

Dohner, J. (2023, March 29). Should you get A guard donkey? Mother Earth News – The Original Guide To Living Wisely; Mother Earth News. https://www.motherearthnews.com/homesteading-and-livestock/guard-donkey-zbcz1310/

Dohner, J. (2023, March 30). Guardian llamas: Pros and cons. Mother Earth News - The Original Guide To Living Wisely; Mother Earth News. https://www.motherearthnews.com/homesteading-and-livestock/guardian-llamas-zbcz1309/

Dohner, J., & Giraud, C. (2018, December 2). Preserving the essential traits and behaviors of livestock guardians do. Jan-Dohner. https://www.jandohner.com/single-post/2018/12/02/preserving-the-essential-traits-and-behaviors-of-livestock-guardian-dogs

Donkey Listener. (2019, July 15). Training donkeys using gentle methods from the whole donkey approach. The Donkey Listener. https://donkeylistener.com/training-donkeys/

Donkey Listener. (2022, May 23). Training a rescue donkey {case study: Gucci}. The Donkey Listener. https://donkeylistener.com/training-a-rescue-donkey/

Exotic Animal Laws by State. (2023). Findlaw.com. https://www.findlaw.com/injury/torts-and-personal-injuries/exotic-animal-laws-by-state.html

Farm safety - handling animals. (n.d.). Gov.au. https://www.betterhealth.vic.gov.au/health/healthyliving/farm-safety-handling-animals

Farmbrite. (2023, January 10). How to Choose the Best Livestock Guardian Dog. Farmbrite. https://www.farmbrite.com/post/how-to-choose-the-best-livestock-guardian-dog

Flint, N. (2022, August 12). The 10 Best Livestock Guardian Dog Breeds, Large and Small. Pet Control HQ. https://petcontrolhq.com/blogs/news/best-livestock-guardian-dog-breeds

Franklin, W.L., N.K. Drufke, and K. J. Powell. 2012. Guard llamas: their use and effectiveness in North America for protecting sheep, goats, cattle, and poultry against canid predators. Pp. 21-36 in: Raggi S., L.A., Rojas S., I., Parraguez G., V.H. and Sepúlveda H., N. (eds.). Resúmenes VI Congreso Mundial Camélidos. Arica, Chile.

Griffler, M. (2019, July 4). How to be safely around A donkey. The Open Sanctuary Project; The Open Sanctuary Project, Inc. https://opensanctuary.org/how-to-be-safely-around-a-donkey/

Griffler, M. (2019, June 14). How donkeys get along with other species. The Open Sanctuary Project; The Open Sanctuary Project, Inc. https://opensanctuary.org/how-donkeys-get-along-with-other-species/

Guard donkey. (2013, May 14). Jacobs Heritage Farm. https://jacobsheritagefarm.com/heritage-livestock/minature-guard-donkey/

Guardian animals for livestock protection and wild dog exclusion. (n.d.). https://pestsmart.org.au/toolkit-resource/guardian-animals-for-livestock-protection-and-wild-dog-exclusion/

Guidelines for using donkeys as guard animals with sheep. (n.d.). Ontario.Ca. https://www.ontario.ca/page/guidelines-using-donkeys-guard-animals-sheep

Guidelines for using donkeys as guard animals with sheep. (n.d.-a). Ontario.Ca. https://www.ontario.ca/page/guidelines-using-donkeys-guard-animals-sheep

Guidelines for using donkeys as guard animals with sheep. (n.d.-b). Ontario.Ca. https://www.ontario.ca/page/guidelines-using-donkeys-guard-animals-sheep

Health and care. (n.d.). Psu.edu. https://extension.psu.edu/animals-and-livestock/llamas-and-alpacas/health-and-care

Heimbuch, J. (2016, June 14). 10 surprising facts about donkeys. Treehugger. https://www.treehugger.com/facts-will-change-way-you-think-about-donkeys-4869321

Introducing predator to prey. (2012, June 28). Horse Canada. https://horse-canada.com/magazine/behaviour/introducing-predator-to-prey/

Kansas Living Magazine. (2016, February 22). Guard donkeys. Kansaslivingmagazine.com. https://kansaslivingmagazine.com/articles/2016/02/22/guard-donkeys

Layers, A. (2013, June 20). Will donkeys keep predators away like foxes and dogs? BackYard Chickens - Learn How to Raise Chickens. https://www.backyardchickens.com/threads/will-donkeys-keep-predators-away-like-foxes-and-dogs.795604/page-2

LeBlanc, T. (2014, June 3). Modern farmer's guide to guard donkeys. Modern Farmer. https://modernfarmer.com/2014/06/modern-farmers-guide-guard-donkeys/

Lee, A. (2020, November 3). 13 reasons why your ranch needs a guard donkey. Helpful Horse Hints. https://www.helpfulhorsehints.com/reasons-to-get-a-guard-donkey/

Lee, A. (2020, October 29). Guard llama or guard donkey: Making the right decision for your farm. Farmhouse Guide; April Lee. https://farmhouseguide.com/guard-llama-vs-guard-donkey/

Lee, A. (2021, December 27). 9 ways donkeys show aggression (and how to stay safe around them). Helpful Horse Hints. https://www.helpfulhorsehints.com/ways-donkeys-show-aggression-and-how-to-stay-safe-around-them/

Liebenberg, L. (n.d.). Self-Rewarding Behaviours in LGD. Blogspot.com. https://predator-friendly-ranching.blogspot.com/2021/05/self-rewarding-behaviours-in-lgd.html

Livestock Guardian Animals. (2023, July 10). Landholders for Dingoes. https://landholdersfordingoes.org/livestock-guardian-animals/

Livestock guardian animals. (n.d.). Wild dog facts. Gov.au. https://www.daf.qld.gov.au/__data/assets/pdf_file/0005/76982/IPA-Wild-Dog-Fact-Sheet-Livestock-Guardian-Dogs.pdf

Livestock guardians. (n.d.). Alpaca Magic. https://www.alpacamagic.com.au/livestock-guardians/

Livestock ownership requirements for small landholders. (n.d.). Gov.au. https://www.agric.wa.gov.au/livestock-biosecurity/livestock-ownership-requirements-small-landholders

Llama. (n.d.). Cosley Zoo. https://cosleyzoo.org/llama/

Llamas and Alpacas. (2018). Govt.Nz. https://www.mpi.govt.nz/dmsdocument/46039-Code-of-Welfare-Layer-llamas-and-alpacas

Mies, C. (2018, April 20). Can dogs live with donkeys? Wagwalking.com; Wag! https://wagwalking.com/sense/can-dogs-live-with-donkey

Multispecies grazing. Country Folks; Lee Newspapers. https://countryfolks.com/multispecies-grazing/

Newspapers, L. (2018, December 11). Animals guarding animals: donkeys and dogs. Country Folks; Lee Newspapers. https://countryfolks.com/animals-guarding-animals-donkeys-and-dogs/

Newspapers, L. (2018, December 11). Animals guarding animals: donkeys and dogs. Country Folks; Lee Newspapers. https://countryfolks.com/animals-guarding-animals-donkeys-and-dogs/

Olson, E. (n.d.). Dogs in Ancient Egypt. DigitalCommons@USU. https://digitalcommons.usu.edu/fsrs2020/66/

Peterson, A. N., Soto, A. P., & McHenry, M. J. (2021). Pursuit and evasion strategies in the predator-prey interactions of fishes. Integrative and Comparative Biology, 61(2), 668–680. https://doi.org/10.1093/icb/icab116

Predation control of livestock. (2014, October 27). Center for Agriculture, Food, and the Environment. https://ag.umass.edu/crops-dairy-livestock-equine/fact-sheets/predation-control-of-livestock

Protection Dogs Worldwide. (2022, September 14). History of Guard Dogs | Protection Dogs Worldwide. https://www.protectiondogs.co.uk/history-of-guard-dogs/

Pugh, B. (n.d.). Chapter 13 Predator Control. Okstate.edu. https://extension.okstate.edu/programs/meat-goat-production/site-files/docs/chapter-13-predator-control.pdf

Purpose, benefits and considerations. (n.d.). Ontario.Ca https://www.ontario.ca/document/livestock-guardian-dogs/purpose-benefits-and-considerations

Quigg, M. (2022, April 5). Llama training tips. Franklinveterinaryclinic.net; Franklin Veterinary Clinic. https://franklinveterinaryclinic.net/llama-training-tips/

Raising llamas & kids. (n.d.). James Skeen. https://www.jamesskeen.com/raising-llamas-and-kids

Shena, T. (2022, January 24). Don't miss the cues with livestock guardian dog behavior. Farm and Dairy. https://www.farmanddairy.com/top-stories/how-to-troubleshoot-livestock-guardian-dog-behavior/700067.html

Sipes, R. (2022a, February 25). Introducing a new donkey to the herd! The Sassy Ass. https://thesassyass.com/blogs/news/introducing-a-new-donkey-to-the-herd

Sipes, R. (2022b, June 27). The basics of donkey training. The Sassy Ass. https://thesassyass.com/blogs/news/the-basics-of-donkey-training

subjecttopressure. (2018, January 27). Correction. Guard Dog Blog. https://guarddogblog.wordpress.com/2018/01/26/correction/

Trollinger, B. (2022, December 12). Beginner's Guide to Livestock Guardian Animals. EcoFarming Daily. https://www.ecofarmingdaily.com/raise-healthy-livestock/beginners-guide-to-livestock-guardian-animals/

Understanding donkey behaviour. (n.d.-a). The Donkey Sanctuary. https://www.thedonkeysanctuary.org.uk/all-about-donkeys/behaviour/understanding-donkey-behaviour

Understanding donkey behaviour. (n.d.-b). The Donkey Sanctuary. https://www.thedonkeysanctuary.org.uk/all-about-donkeys/behaviour/understanding-donkey-behaviour

Understanding the instincts of livestock guardian dogs. (2023, August 12). Off Leash Blog. https://blog.tryfi.com/livestock-guardian-dogs/

Usda Aphis. (n.d.). Usda.gov. https://www.aphis.usda.gov/aphis/ourfocus/wildlifedamage/operational-activities/sa_livestock/ct_protecting_livestock_predators

User, S. (2020, December 21). The livestock guardian dog: The best friend of livestock in extensive systems. Fawec.Org. https://www.fawec.org/en/what-do-we-do/inspiring-pilot-farms/363-le-chien-de-protection-de-troupeaux-le-meilleur-ami-de-l-elevage-pastoral

Using llamas as guardians: Benefits and considerations. (2023, May 15). The Thrifty Homesteader. https://thriftyhomesteader.com/llamas-as-guardians/

Verana, C. (2023, April 4). 15 donkey facts about the misunderstood equines. TRVST. https://www.trvst.world/biodiversity/donkey-facts/

Vistein, G. (2016, July 9). Donkeys as Guardians of your livestock. Farming with Carnivores Network. https://farmingwithcarnivoresnetwork.com/donkeys-guardians-livestock/

Weaver, S. (2016, March 31). Choosing a livestock guardian. Grit - Rural American Know-How. https://www.grit.com/animals/livestock/livestock-guardians-ze0z1603zcbru/

Weaver, S. (2019, April 26). Livestock guardian Donkeys. Grit - Rural American Know-How. https://www.grit.com/animals/livestock-guardian-donkeys-ze0z1904znad/

Wildlife Specialist. (n.d.). How LGD reduces predation. Tamu.edu. https://sanangelo.tamu.edu/files/2013/08/Livestock-Guardian-Dogs1.pdf

Wyzard, B. (2020a, September 9). Livestock Guardian Dogs and Food: What to Feed, When, and Problems to Avoid - For Love of Livestock. For Love of Livestock. https://www.forloveoflivestock.com/blog/livestock-guardian-dogs-and-food

Wyzard, B. (2020b, September 10). Training Livestock Guardian Dogs: The Ultimate Guide - For Love of Livestock. For Love of Livestock. https://www.forloveoflivestock.com/blog/training-livestock-guardian-dogs-the-ultimate-guide

Yokhna, D. (2009, February 18). Protect your flock with guard donkeys. Hobby Farms. https://www.hobbyfarms.com/protect-your-flock-with-guard-donkeys-2/

www.ingramcontent.com/pod-product-compliance
Lightning Source LLC
Chambersburg PA
CBHW070150310326
41914CB00089B/758